RSGB DELUXE LOG BOOK & DIARY

2018

Contents

Published by the Radio Society of Great Britain 3 Abbey Court, Priory Business Park, Bedford, MK44 3WH

Publisher's note
While the information presented is believed to be correct, the authors, the publisher and their agents cannot accept responsibility for consequences arising from any inaccuracies or omissions.

Production & DTP : Mark Allgar, M1MPA
Cover Design: Kevin Williams, M6CYB
Printed in Great Britain by Page Bros Ltd of Norwich

RSGB Band Plan

The following band plan is largely based on that agreed at IARU Region 1 General Conferences with some local differences on frequencies above 430MHz.

HF

The addition of a usage note in the 472kHz band and a wider bandwidth all-modes segment in 29.0-29.1MHz that were agreed at Varna.

VHF/UHF

The most noticeable feature is that 146-147MHz has been included. However IARU changes and the Ofcom-ETCC packet review also result in changes to the main 145MHz band. Several packet channels have been cleared whilst the bottom of the band is now shared with new narrowband amateur satellite downlinks.

Both 145MHz and other VHF bands see the deletion of old RTTY and FAX channels making room for all-modes usage. 432-433MHz also sees some change including a more consistent designation for the 12.5kHz operation of Internet gateways. A landmark change is the removal of the UK beacon segment, and IARU beacon frequencies.

MICROWAVE

The Ofcom spectrum release changes see the 2350-2390 and 3410-3475MHz ranges removed from the appropriate band plans and some of the remaining frequencies being reset to all modes. In future this may change further as new data and DATV developments become clear. The new 2300-2302MHz segment (if you have the NoV) is incorporated as a separate table. The 10GHz band also sees a clearout of old designations and updates for repeater and wideband usage. A new shaded warning zone in the bottom 1010.125GHz section indicates where the Primary User now has increased use, having been pressured out of other spectrum.

GENERAL NOTES

These have also been updated, including the need to refer to certain bands. New notes provide usage. Another new note, agreed by IARU Region 1 at Varna emphasises that all VHF WSPR frequencies in the band plans are transmitted centre frequencies and not ambiguous dial settings.

FINALLY…

As we have said before, band plans are living entities and do evolve over time. Please ensure you only refer or link to the current ones on the RSGB website and remove any older ones you have locally.

The band plan including the master section are on the RSGB website – and if you are unsure, by all means contact the HF, VHF or Microwave Spectrum Manager via:

hf.manager@rsgb.org.uk or

vhf.manager@rsgb.org.uk or

mw.manager@rsgb.org.uk

136kHz	NECESSARY BANDWIDTH	UK USAGE
135.7-137.8kHz	200Hz	CW, QRSS and Narrowband Digital Modes

Licence Notes: Amateur Service – Secondary User. 1 watt (0dBW) ERP.
R.R. 5.67B. The use of the band 135.7-137.8kHz in Algeria, Egypt, Iran (Islamic Republic of), Iraq, Lebanon, Syrian Arab Republic Sudan, South Sudan and Tunisia is limited to fixed and maritime mobile services. The amateur service shall not be used in the above-mentioned countries in the band 135.7-137.8kHz, and this should be taken into account by the countries authorising such use. (WRC-12).

472kHz (600m)	NECESSARY BANDWIDTH	UK USAGE

IARU Region 1 does not have a formal band plan for this allocation but has a usage recommendation (Note 1).

| 472-479kHz | 500Hz | CW, QRSS and Narrowband Digital Modes |

Note 1: Usage recommendation – 472-475kHz CW only 200Hz maximum bandwidth, 475-479kHz CW and Digimodes.
Note 2: It should be emphasised that this band is available on a non-interference basis to existing services. UK amateurs should be aware that some overseas stations may be restricted in terms of transmit frequency in order to avoid interference to nearby radio navigation service Non-Directional Beacons.
Licence Notes: Amateur Service – Secondary User. Full Licensees only, **5 watts EIRP maximum**. Note that conditions regarding this band are specified by the Licence Schedule notes.
R.R. 5.80B. The use of the frequency band 472-479kHz in Algeria, Saudi Arabia, Azerbaijan, Bahrain, Belarus, China, Comoros, Djibouti, Egypt, United Arab Emirates, the Russian Federation, Iraq, Jordan, Kazakhstan, Kuwait, Lebanon, Libya, Mauritania, Oman, Uzbekistan, Qatar, Syrian Arab Republic, Kyrgyzstan, Somalia, Sudan, Tunisia and Yemen is limited to the maritime mobile and aeronautical radionavigation services. The amateur service shall not be used in the above-mentioned countries in this frequency band, and this should be taken into account by the countries authorising such use. (WRC 12).

1.8MHz (160m)	NECESSARY BANDWIDTH	UK USAGE
1,810-1,838kHz	200Hz	Telegraphy
1,838-1,840	500Hz	Narrowband Modes
1,840-1,843	2.7kHz	All Modes
1,843-2,000	2.7kHz	Telephony (Note 1), Telegraphy 1,836kHz – QRP (low power) Centre of Activity 1,960kHz – DF Contest Beacons (14dBW)

Note 1: Lowest LSB carrier frequency (dial setting) should be 1,843kHz. AX25 packet should not be used on the 1.8MHz band.
Licence Notes: 1,810-1,850kHz – Primary User: 1,810-1,830kHz on a non-interference basis to stations outside of the UK. 1,850-2,000kHz – Secondary User. 32W (15dBW) maximum.
Notes to the Band Plan: As on page 167.

3.5MHz (80m)	NECESSARY BANDWIDTH	UK USAGE
3,500-3,510kHz	200Hz	Telegraphy – Priority for Inter-Continental Operation
3,510-3,560	200Hz	Telegraphy – Contest Preferred. 3,555kHz – QRS (slow telegraphy) Centre of Activity
3,560-3,570	200Hz	Telegraphy 3,560kHz – QRP (low power) Centre of Activity
3,570-3,580	200Hz	Narrowband Modes
3,580-3,590	500Hz	Narrowband Modes
3,590-3,600	500Hz	Narrowband Modes – Automatically Controlled Data Stations (unattended)
3,600-3,620	2.7kHz	All Modes – Automatically Controlled Data Stations (unattended), (Note 1)
3,600-3,650	2.7kHz	All Modes – Phone Contest Preferred, (Note 1). 3,630kHz – Digital Voice Centre of Activity
3,650-3,700	2.7kHz	All Modes – Telephony, Telegraphy 3,663kHz May Be Used For UK Emergency Comms Traffic 3,690kHz SSB QRP (low power) Centre of Activity
3,700-3,775	2.7kHz	All Modes – Phone Contest Preferred 3,735kHz – Image Mode Centre of Activity 3,760kHz – IARU Region 1 Emergency Centre of Activity
3,775-3,800	2.7kHz	All modes - Phone contest preferred Priority for Inter-Continental Telephony (SSB) Operation

Note 1. Lowest LSB carrier frequency (dial setting) should be 3,603kHz.
Licence Notes: Primary User: Shared with other user services.
Notes to the Band Plan: As on page 167.

5MHz (60m)	AVAILABLE WIDTH	UK USAGE
5,258.5-5,264kHz	5.5kHz	5,262kHz – CW QRP Centre of Activity
5,276-5,284	8kHz	5,278.5kHz – May be used for UK Emergency Comms Traffic
5,288.5-5,292	3.5kHz	Beacons on 5290kHz (Note 2), WSPR
5,298-5,307	9kHz	
5,313-5,323	10kHz	5,317kHz – AM 6kHz maximum bandwidth
5,333-5,338	5kHz	
5,354-5,358	4kHz	
5,362-5,374.5	12.5kHz	5,362-5,370kHz – Digital Mode Activity in the UK
5,378-5,382	4kHz	
5,395-5,401.5	6.5kHz	
5,403.5-5,406.5	3kHz	5,403.5kHz – USB Common International Frequency

Unless indicated, usage is All Modes (necessary bandwidth to be within channel limits).
Note 1: Upper Sideband is recommended for SSB activity.
Note 2: Activity should avoid interference to the experimental beacons on 5290kHz.
Note 3: Amplitude Modulation is permitted with a maximum bandwidth of 6kHz, on frequencies with at least 6kHz available width.
Note 4: Contacts within the UK should avoid the WRC-15 allocation if possible
Licence Notes: Full Licensees only, **Secondary User, 100 watts maximum.** Note that conditions on transmission bandwidth, power and antennas are specified in the Licence.
Notes to the Band Plan. As on page 167.

7MHz (40m)	NECESSARY BANDWIDTH	UK USAGE
7,000-7,040kHz	200Hz	Telegraphy – 7,030kHz QRP (low power) Centre of Activity
7,040-7,047	500Hz	Narrowband Modes (Note 2)
7,047-7,050	500Hz	Narrowband Modes, Automatically Controlled Data Stations (unattended)
7,050-7,053	2.7kHz	All Modes, Automatically Controlled Data Stations (unattended), (Note 1)
7,053-7,060	2.7kHz	All Modes, Digimodes
7,060-7,100	2.7kHz	All Modes, SSB Contest Preferred Segment Digital Voice 7,070kHz; SSB QRP Centre of Activity 7,090kHz
7,100-7,130	2.7kHz	All Modes, 7,110kHz – Region 1 Emergency Centre of Activity
7,130-7,200	2.7kHz	All Modes, SSB Contest Preferred Segment; 7,165kHz – Image Centre of Activity
7,175-7,200	2.7kHz	All Modes, Priority For Inter-Continental Operation

Note 1: Lowest LSB carrier frequency (dial setting) should be 7,053kHz.
Note 2: PSK31 activity starts from 7,040kHz. Since 2009, the narrowband modes segment starts at 7,040kHz.
Licence Notes: 7,000-7,100kHz Amateur and Amateur Satellite Service – Primary User.
7,100-7,200kHz Amateur Service – Primary User.
Notes to the Band Plan. As on page 167.

10MHz (30m)	NECESSARY BANDWIDTH	UK USAGE
10,100-10,130kHz	200Hz	Telegraphy (CW)
		10,116kHz – QRP (low power) Centre of Activity
10,130-10,150	500Hz	Narrowband Modes
		Automatically Controlled Data Stations (unattended) should avoid the use of the 10MHz band

Licence Notes: Amateur Service – Secondary User.
Notes to the Band Plan: As on page 167.
The 10MHz band is allocated to the amateur service only on a secondary basis. The IARU has agreed that only CW and other narrow bandwidth modes are to be used on this band. Likewise the band is not to be used for contests and bulletins. SSB may be used on the 10MHz band during emergencies involving the immediate safety of life and property, and only by stations actually involved with the handling of emergency traffic. The band segment 10,120-10,140kHz may only be used for SSB transmissions in the area of Africa south of the equator during local daylight hours.

14MHz (20m)	NECESSARY BANDWIDTH	UK USAGE
14,000-14,060kHz	200Hz	Telegraphy – Contest Preferred
		14,055kHz – QRS (slow telegraphy) Centre of Activity
14,060-14,070	200Hz	Telegraphy
		14,060kHz – QRP (low power) Centre of Activity
14,070-14,089	500Hz	Narrowband Modes
14,089-14,099	500Hz	Narrowband Modes – Automatically Controlled Data Stations (unattended)
14,099-14,101		IBP – Reserved Exclusively for Beacons
14,101-14,112	2.7kHz	All Modes – Automatically Controlled Data Stations (unattended)
14,112-14,125	2.7kHz	All Modes (excluding digimodes)
14,125-14,300	2.7kHz	All Modes – SSB Contest Preferred Segment
		14,130kHz – Digital Voice Centre of Activity
		14,195 ±5kHz – Priority for DXpeditions
		14,230kHz – Image Centre of Activity
		14,285kHz – QRP Centre of Activity
14,300-14,350	2.7kHz	All Modes
		14,300kHz – Global Emergency Centre of Activity

Licence Notes: Amateur Service – Primary User. 14,000-14,250kHz Amateur Satellite Service – Primary User.
Notes to the Band Plan: As on page 167.

18MHz (17m)	NECESSARY BANDWIDTH	UK USAGE
18,068-18,095kHz	200Hz	Telegraphy – 18,086kHz QRP (low power) Centre of Activity
18,095-18,105	500Hz	Narrowband Modes
18,105-18,109	500Hz	Narrowband Modes – Automatically Controlled Data
18,111-18,120	2.7kHz	All Modes – Automatically Controlled Data Stations (unattended)
18,120-18,168	2.7kHz	All Modes, 18,130kHz – SSB QRP Centre of Activity
		18,150kHz – Digital Voice Centre of Activity
		18,160kHz – Global Emergency Centre of Activity

Licence Notes: Amateur and Amateur Satellite Service – Primary User. The band is not to be used for contests or bulletins.
Notes to the Band Plan: As on page 167.

21MHz (15m)	NECESSARY BANDWIDTH	UK USAGE
21,000-21,070kHz	200Hz	Telegraphy
		21,055kHz – QRS (slow telegraphy) Centre of Activity
		21,060kHz – QRP (low power) Centre of Activity
21,070-21,090	500Hz	Narrowband Modes
21,090-21,110	500Hz	Narrowband Modes – Automatically Controlled Data Stations (unattended)
21,110-21,120	2.7kHz	All Modes (excluding SSB) – Automatically Controlled Data Stations (unattended)
21,120-21,149	500Hz	Narrowband Modes
21,149-21,151		IBP – Reserved Exclusively For Beacons
21,151-21,450	2.7kHz	All Modes
		21,180kHz – Digital Voice Centre of Activity
		21,285kHz – QRP Centre of Activity
		21,340kHz – Image Centre of Activity
		21,360kHz – Global Emergency Centre of Activity

Licence Notes: Amateur and Amateur Satellite Service – Primary User.
Notes to the Band Plan: As on page 167.

24MHz (12m)	NECESSARY BANDWIDTH	UK USAGE
24,890-24,915kHz	200Hz	Telegraphy
		24,906kHz – QRP (low power) Centre of Activity
24,915-24,925	500Hz	Narrowband Modes
24,925-24,929	500Hz	Narrowband Modes – Automatically Controlled Data Stations (unattended)
24,929-24,931		IBP – Reserved Exclusively For Beacons
24,931-24,940	2.7kHz	All Modes – Automatically Controlled Data Stations (unattended)
24,940-24,990	2.7kHz	All Modes, 24,950kHz – SSB QRP Centre of Activity
		24,960kHz – Digital Voice Centre of Activity

Licence Notes: Amateur and Amateur Satellite Service – Primary User. The band is not to be used for contests or bulletins.
Notes to the Band Plan: As on page 167.

28MHz (10m)	NECESSARY BANDWIDTH	UK USAGE
28,000-28,070kHz	200Hz	Telegraphy
		28,055kHz – QRS (slow telegraphy) Centre of Activity
		28,060kHz – QRP (low power) Centre of Activity
28,070-28,120	500Hz	Narrowband Modes
28,120-28,150	500Hz	Narrowband Modes – Automatically Controlled Data Stations (unattended)
28,150-28,190	500Hz	Narrowband Modes
28,190-28,199		IBP – Regional Time Shared Beacons
28,199-28,201		IBP – World Wide Time Shared Beacons
28,201-28,225		IBP – Continuous-Duty Beacons
28,225-28,300	2.7kHz	All Modes – Beacons
28,300-28,320	2.7kHz	All Modes – Automatically Controlled Data Stations (unattended)
28,320-29,000	2.7kHz	All modes
		28,330kHz – Digital Voice Centre of Activity
		28,360kHz – QRP Centre of Activity
		28,680kHz – Image Centre of Activity
29,000-29,100	6kHz	All Modes
29,100-29,200	6kHz	All Modes – FM Simplex – 10kHz Channels
29,200-29,300	6kHz	All Modes – Automatically Controlled Data Stations (unattended)
		29,270kHz – Internet Gateways Channel
		29,280kHz – UK Internet Voice Gateway (unattended)
		29,290kHz – UK Internet Voice Gateway (unattended)
29,300-29,510	6kHz	Satellite Links
29,510-29,520		Guard Channel
29,520-29,590	6kHz	All Modes – FM Repeater Inputs (RH1-RH8)
29,600	6kHz	All Modes – FM Calling Channel
29,610	6kHz	All Modes – FM Simplex Repeater (parrot) – input and output
29,620-29,700	6kHz	All Modes – FM Repeater Outputs (RH1-RH8)

Licence Notes: Amateur and Amateur Satellite Service – Primary User: 26dBW permitted. Beacons may be established for DF competitions except within 50km of NGR SK985640 (Waddington).
Notes to the Band Plan: As on page 167.

50MHz (6m)	NECESSARY BANDWIDTH	UK USAGE
50.000-50.100MHz	500Hz	Telegraphy Only (except for Beacon Project) (Note 2)
		50.000-50.030MHz reserved for future Synchronised Beacon Project (Note 2)
		Region 1: 50.000-50.010; Region 2: 50.010-50.020; Region 3: 50.020-50.030
		50.050MHz – Future International Centre of Activity
		50.090MHz – Inter-Continental DX Centre of Activity (Note 1)
50.100-50.200	2.7kHz	SSB/Telegraphy – International Preferred
		50.100-50.130MHz – Inter-Continental DX Telegraphy & SSB (Note 1)
		50.110MHz – Inter-Continental DX Centre of Activity
		50.130-50.200MHz – General International Telegraphy & SSB
		50.150MHz – International Centre of Activity
50.200-50.300	2.7kHz	SSB/Telegraphy – General Usage
		50.285MHz – Crossband Centre of Activity
50.300-50.400	2.7kHz	MGM/Narrowband/Telegraphy
		50.305MHz – PSK Centre of Activity
		50.310-50.320MHz – EME
		50.320-50.380MHz – MS
50.400-50.500		**Propagation Beacons only**
50.500-52.000	12.5kHz	All Modes
		50.510MHz – SSTV (AFSK)
		50.520MHz – Internet Voice Gateway (10kHz channels), (IARU common channel)
		50.530MHz – Internet Voice Gateway (10kHz channels), (IARU common channel)
		50.540MHz – Internet Voice Gateway (10kHz channels), (IARU common channel)
		50.550MHz – Image/Fax working frequency
		50.600MHz – RTTY (FSK)
		50.620-50.750MHz – Digital communications
		50.630MHz – Digital Voice (DV) calling
		50.710-50.890MHz – FM/DV Repeater Outputs (10kHz channel spacing)
		51.210-51.390MHz – FM/DV Repeater Inputs (10kHz channel spacing) (Note 4)
		51.410-51.590MHz – FM/DV Simplex (Note 3) (Note 4)
		51.510MHz – FM Calling Frequency
		51.530MHz – GB2RS News Broadcast and Slow Morse
		51.650 & 51.750MHz – See Note 5 (25kHz aligned)
		51.770 & 51.790MHz – See Note 5
		51.810-51.990MHz – FM/DV Repeater Outputs (IARU aligned channels)

Note 1: Only to be used between stations in different continents (not for intra-European QSOs).
Note 2: 50.0-50.1MHz is currently shared with Propagation Beacons. These are due to be migrated by Aug 2014 to 50.4-50.5MHz, to create more space for Telegraphy and a new Synchronised Beacon Project.
Note 3: 20kHz channel spacing. Channel centre frequencies start at 51.430MHz.
Note 4: Embedded data traffic is allowed with digital voice (DV).
Note 5: May be used for Emergency Communications and Community Events.
Licence Notes: Amateur Service 50.0-51.0MHz – Primary User. Amateur Service 51.0-52.0MHz – Secondary User. 100W (20dBW) maximum. Available on the basis on non-interference to other services (inside or outside the UK).
Notes to the Band Plan: As on page 167.

70MHz (4m)	NECESSARY BANDWIDTH	UK USAGE (NOTE 1)
70.000-70.090MHz	**1kHz**	**Propagation Beacons Only**
70.090-70.100	1kHz	Personal Beacons
70.100-70.250	2.7kHz	Narrowband Modes
		70.185MHz – Cross-band Activity Centre
		70.200MHz – CW/SSB Calling
		70.250MHz – MS Calling
70.250-70.294	12kHz	All Modes
		70.260MHz – AM/FM Calling
		70.270MHz MGM Centre of Activity
70.294-70.500	12kHz	All Modes Channelised Operations Using 12.5kHz Spacing
		70.3000MHz
		70.3125MHz – Digital Modes
		70.3250MHz – DX Cluster
		70.3375MHz – Digital Modes
		70.3500MHz – Internet Voice Gateway (Note 2)
		70.3625MHz – Internet Voice Gateway
		70.3750MHz – See Note 2
		70.3875MHz – Internet Voice Gateway
		70.4000MHz – See Note 2
		70.4125MHz – Internet Voice Gateway
		70.4250MHz – FM Simplex – used by GB2RS news broadcast
		70.4375MHz – Digital Modes (special projects)
		70.4500MHz – FM Calling
		70.4625MHz – Digital Modes
		70.4750MHz
		70.4875MHz – Digital Modes

Note 1: Usage by operators in other countries may be influenced by restrictions in their national allocations.
Note 2: May be used for Emergency Communications and Community Events.
Licence Notes: Amateur Service 70.0-70.5MHz – Secondary User: 160W (22dBW) maximum. Available on the basis of non-interference to other services (inside or outside the UK).
Notes to the Band Plan: As on page 167.

144MHz (2m)	NECESSARY BANDWIDTH	UK USAGE
144.000-144.025MHz	2700Hz	All Modes – including Satellite Downlinks
144.025-144.110	500Hz	Telegraphy (including EME CW)
		144.050MHz – Telegraphy Centre of Activity
		144.100MHz – Random MS Telegraphy Calling, (Note 1)
144.110-144.150	500Hz	Telegraphy and MGM
		144.138MHz – PSK31 Centre of Activity
		EME MGM Activity (Note 7)
144.150-144.180	2700Hz	Telegraphy, MGM and SSB
144.180-144.360	2700Hz	Telegraphy and SSB
		144.175MHz – Microwave Talk-back
		144.195-144.205MHz – Random MS SSB
		144.200MHz – Random MS SSB Calling Frequency
		144.250MHz – GB2RS News Broadcast and Slow Morse
		144.260MHz – USB. (Note 10)
		144.300MHz – SSB Centre of Activity
144.360-144.399	2700Hz	Telegraphy, MGM, SSB
		144.370MHz – MGM Calling Frequency
144.400-144.490		**Propagation Beacons only**
144.490-144.500		Beacon guard band
144.500-144.794	20kHz	All Modes (Note 8)
		144.500MHz – Image Modes Centre (SSTV, FAX, etc)
		144.600MHz – Data Centre of Activity (MGM, RTTY, etc)
		144.6125MHz – UK Digital Voice (DV) Calling (Note 9)
		144.625-144.675MHz – See Note 10
		144.750MHz – ATV Talk-back
		144.775-144.794MHz – See Note 10
144.794-144.990	12kHz	MGM Digital Communications (Note 15)
		144.800-144.9875MHz – MGM/Digital Communications
		144.8000MHz – Unconnected Nets – APRS, UiView etc (Note 14)
		144.8125MHz – DV Internet Voice Gateway
		144.8250MHz – DV Internet Voice Gateway
		144.8375MHz – DV Internet Voice Gateway
		144.8500MHz – DV Internet Voice Gateway
		144.8625MHz – DV Internet Voice Gateway
		144.9250MHz – TCP/IP Usage
		144.9375MHz – AX25 Usage
		144.9500MHz – AX25 Usage
		144.9625MHz – FM Internet Voice Gateway
		144.9750MHz, 144.9875MHz To Be Decided (Note 11)
144.990-145.1935	12kHz	FM/DV RV48-RV63 Repeater Input Exclusive (Note 2 & 5)
145.200	12kHz	FM/DV Space Communications (eg ISS) – Earth-to-Space
		145.2000MHz – (Note 4 & 10)
145.200-145.5935	12kHz	FM/DV V16-V48 – FM/DV Simplex (Note 3, 5 & 6)
		145.2250MHz – See Note 10
		145.2375MHz – FM Internet Voice Gateway (IARU common channel)
		145.2500MHz – Used for Slow Morse Transmissions
		145.2875MHz – FM Internet Voice Gateway (IARU common channel)
		145.3375MHz – FM Internet Voice Gateway (IARU common channel)
		145.5000MHz – FM Calling (Note 12)
		145.5250MHz – Used for GB2RS News Broadcast.
		145.5500MHz – Used for Rally/exhibition Talk-in
		145.5750MHz, 145.5875MHz (Note 11)
145.5935-145.7935	12kHz	FM/DV RV48-RV63 – Repeater Output (Note 2)
145.800	12kHz	FM/DV Space Communications (eg ISS) – Space-Earth
145.806-146.000	12kHz	All Modes – Satellite Exclusive

Note 1: Meteor scatter operation can take place up to 26kHz higher than the reference frequency.
Note 2: 12.5kHz channels numbered RV48-RV63. RV48 input = 145.000MHz, output = 145.600MHz.
Note 3: 12.5kHz simplex channels numbered V16-V46. V16 = 145.200MHz.
Note 4: Emergency Communications Groups utilising this frequency should take steps to avoid interference to ISS operations in non-emergency situations.
Note 5: Embedded data traffic is allowed with digital voice (DV).
Note 6: Simplex use only – no DV gateways.
Note 7: EME activity using MGM is commonly practiced between 144.110-144.160MHz.
Note 8: Amplitude Modulation (AM) is acceptable within the All Modes segment. AM usage is typically found on 144.550MHz. Users should consider adjacent channel activity when selecting operating frequencies.
Note 9: In other countries IARU Region 1 recommends 145.375MHz.
Note 10: May be used for Emergency Communications and Community Events.
Note 11: May be used for repeaters in other IARU Region 1 countries.
Note 12: DV users are asked not to use this channel, and use 144.6125MHz for calling.
Note 13: Not used.
Note 14: 144.800 use should be NBFM to avoid interference to 144.8125 DV Gateways.
Licence Notes: Amateur Service and Amateur Satellite Service – Primary User. Beacons may be established for DF competitions except within 50km of TA 012869 (Scarborough).
Notes to the Band Plan: As on page 167.

146MHz

146MHz	NECESSARY BANDWIDTH	UK USAGE
146.000-146.900MHz	500kHz	Wideband Digital Modes (High speed data, DATV etc)
		146.500MHz Centre frequency for wideband modes (Note 1)
146.900-147.000MHz	12kHz	Narrowband Digital Modes including Digital Voice
		146.900
		146.9125
		146.925
		146.9375 Not available in/near Scotland (see Licence Notes & NoV terms)
		146.9500
		146.9625
		146.9750
		146.9875

Note 1: Users of wideband modes must ensure their spectral emissions are contained with the band limits.

Licence Notes: Full Licensees only, with NoV, 25W ERP max – not available in the Isle of Man or Channel Isles. Note that additional restrictions on geographic location, antenna height and upper frequency limit are specified by the NoV terms.

It should be emphasised that this band is UK-specific and is available on a non-interference basis to existing services. Upper Band limit 147.000MHz (or 146.93750 where applicable) are absolute limits and not centre frequencies. The absolute band frequency limit in or within 40km of Scotland is 146.93750MHz – see NoV schedule

Notes to the Band Plan: As on page 167.

430MHz (70cm)

430MHz (70cm) IARU Recommendation	NECESSARY BANDWIDTH	UK USAGE
430.0000-431.9810MHz	20kHz	430.0125-430.0750MHz – FM Internet Voice Gateways (Notes 7, 8)
All Modes		
		430.4000-430.7750 – UK DV 9MHz Split Repeaters – inputs
Digital Links		
430.6000-430.9250		430.8000MHz – 7.6MHz Talk-through (Note 10)
Digital Repeaters		430.8250-430.9750MHz – RU66-RU78 7.6MHz Split Repeaters – outputs
		See Licence Exclusion Note; 431-432MHz
		430.9900-431.9000MHz – Digital Communications
		431.0750-431.1750MHz – DV Internet Voice Gateways (Note 8)
432.0000-432.1000	500Hz	432.0000-432.0250MHz – Moonbounce (EME)
Telegraphy		432.0500MHz – Telegraphy Centre of Activity
MGM		432.0880MHz – PSK31 Centre of Activity
432.1000-432.4000	2700Hz	432.2000MHz – SSB Centre of Activity
SSB, Telegraphy		432.3500MHz – Microwave Talk-back (Europe)
MGM		432.3700MHz – FSK441 Calling Frequency
432.4000-432.5000	500Hz	Propagation Beacons only
Beacons Exclusive		
432.5000-432.9940	25kHz	432.5000MHz – Narrowband SSTV Activity
Centre		
All Modes	(Note 11)	432.6250-432.6750MHz Digital Communications (25kHz channels)
Non-channelised		432.7750MHz 1.6MHz Talk-through – Base TX (Note 10)
432.9940-433.3810	25kHz	433.0000-433.3750MHz (RB0-RB15) – RU240-RU270
FM repeater outputs in UK only (Note 1)	(Note 11)	FM/DV Repeater Outputs (25kHz channels) in UK Only
433.3940-433.5810	25kHz	433.4000MHz U272 – IARU Region 1 SSTV (FM/AFSK)
	(Note 11)	433.4250MHz U274
FM/DV (Notes 12, 13)		433.450MHz U276 (Note 5)
Simplex Channels		433.4750MHz U278
		433.5000MHz U280 – FM Calling Channel
		433.5250MHz U282
		433.5500MHz U284 – Used for Rally/Exhibition Talk-in
		433.5750MHz U286
433.6000-434.0000	25kHz	433.6250-6750MHz – Digital Communications (25kHz channels)
All Modes	(Note 11)	
433.800MHz for		433.700MHz (Note 10)
APRS where 144.800MHz cannot be used		433.7250-433.7750MHz (Note 10)
		433.8000-434.2500MHz – Digital Communications
434.000-434.5940	25kHz	433.9500-434.0500MHz – Internet Voice Gateways (Note 8)
	(Note 11)	434.3750MHz 1.6MHz Talk-through – Mobile TX (Note 10)
		434.4750-434.5250MHz – Internet Voice Gateways (Note 8)
434.5940-434.9810	25kHz	434.6000-434.9750MHz (RB0-RB15) RU240-RU270
FM repeater inputs in UK only & ATV (Note 4)	(Note 11)	FM/DV Repeater Inputs (25kHz channels) in UK Only (Note 12)
435.0000-438.0000	20kHz	Satellites and Fast Scan TV (Note 4)
		437.0000 – Experimental DATV Centre of Activity (Note 14)
438.0000-440.0000	25kHz	438.0250-438.1750MHz – IARU Region 1 Digital Communications
All Modes	(Note 11)	438.2000-439.4250MHz (Note 1)
		438.4000MHz – 7.6MHz Talk-through (Note 10)
		438.4250-438.5750MHz RU66-RU78 – 7.6MHz Split Repeaters – inputs

430MHz (70cm) IARU Recommendation	NECESSARY BANDWIDTH	UK USAGE Contd
		438.6125MHz – UK DV calling (Note 12) (Note 13)
		439.6000-440.0000MHz – Digital Communications
		439.400-439.775MHz – UK DV 9MHz split repeaters - Outputs

Note 1: In Switzerland, Germany and Austria, repeater inputs are 431.050-431.825MHz with 25kHz spacing and outputs 438.650-439.425MHz. In Belgium, France and the Netherlands repeater outputs are 430.025-430.375MHz with 12.5kHz spacing and inputs at 431.625-431.975MHz. In other European countries repeater inputs are 433.000-433.375MHz with 25kHz spacing and outputs at 434.600-434.975MHz, ie the reverse of the UK allocation.

Note 4: ATV carrier frequencies shall be chosen to avoid interference to other users, in particular the satellite service and repeater inputs.

Note 5: In other countries IARU Region 1 recommends 433.450MHz for DV calling.

Note 7: Users must accept interference from repeater output channels in France and the Netherlands at 430.025-430.575MHz. Users with sites that allow propagation to other countries (notably France and the Netherlands) must survey the proposed frequency before use to ensure that they will not cause interference to users in those countries.

Note 8: All internet voice gateways: 12.5kHz channels, maximum deviation ±2.4kHz, maximum effective radiated power 5W (7dBW), attended only operation in the presence of the NoV holder.

Note 10: May be used for Emergency Communications and Community Events.

Note 11: IARU Region 1 recommended maximum bandwidths are 12.5 or 20kHz.

Note 12: Embedded data traffic is allowed with digital voice (DV).

Note 13: Simplex use only - no DV gateways.

Note 14: QPSK 2 Mega-symbols/second maximum recommended.

Licence Notes: Amateur Service – Secondary User. Amateur Satellite Service: 435-438MHz – Secondary User. Exclusion: 431-432MHz not available within 100km radius of Charing Cross, London. Power Restriction 430-432MHz is 40 watts effective radiated power maximum.

Notes to the Band Plan: As on page 167.

1.3GHz (23cm)

1.3GHz (23cm)	NECESSARY BANDWIDTH	UK USAGE
1240.000-1240.500MHz	2700Hz	Alternative Narrowband Segment – see Note 7 – 1240.00-1240.750MHz
1240.500-1240.750		Alternative Propagation Beacon Segment
1240.750-1241.000	20kHz	FM/DV Repeater Inputs
1241.000-1241.750	150kHz	DD High Speed Digital Data – 5 x 150kHz channels
All Modes		1241.075, 1241.225, 1241.375, 1241.525, 1241.675MHz (±75kHz)
1241.750-1242.000	20kHz	25kHz Channels available for FM/DV use
All Modes		1241.775-1241.975MHz
1242.000-1249.000		TV Repeaters (Note 9)
ATV		New DATV Repeater Inputs
		Original ATV Repeater Inputs: 1248, 1249
1249.000-1249.250	20kHz	FM/DV Repeater Outputs, 25kHz Channels (Note 9)
		1249.025-1249.225MHz
1250.00		In order to prevent interference to Primary Users, caution must be exercised prior to using 1250-1290MHz in the UK
1260.000-1270.000		Amateur Satellite Service – Earth to Space Uplinks Only
Satellites		
1290.000		
1290.994-1291.481	20kHz	FM/DV Repeater Inputs (Note 5)
		1291.000-1291.375MHz (RM0-RM15) 25kHz spacing
1291.494-1296.000	All Modes	
All Modes		Preferred Narrowband segment
1296.000-1296.150	500Hz	1296.000-1296.025MHz – Moonbounce
Telegraphy, MGM		1296.138MHz – PSK31 Centre of Activity
1296.150-1296.800	2700Hz	1296.200MHz – Narrowband Centre of Activity
Telegraphy, SSB & MGM		1296.400-1296.600MHz – Linear Transponder Input
(Note 1)		1296.500MHz – Image Mode Centre of Activity (SSTV, FAX etc)
		1296.600MHz – Narrowband Data Centre of Activity (MGM, RTTY etc)
		1296.600-1296.700MHz – Linear Transponder Output
		1296.750-1296.800MHz – Local Beacons, 10W ERP max
1296.800-1296.994		1296.800-1296.990MHz – Propagation Beacons only
		Beacons exclusive
1296.994-1297.481	20kHz	FM/DV Repeater Outputs (Note 5)
		1297.000-1297.375MHz (RM0-RM15)
1297.494-1297.981	20kHz	FM/DV Simplex ((Notes 2, 5 & 6)) 25kHz spacing
		1297.500-1297.750MHz (SM20-SM30)
FM/DV simplex		1297.725MHz – Digital Voice (DV) Calling (IARU recommended)
(Notes 2, 5, 6)		1297.900-1297.975MHz – FM Internet Voice Gateways (IARU common channels, 25kHz)
1298.000-1299.000	20kHz	All Modes
All Modes		General mixed analogue or digital use in channels
		1298.025-1298.975MHz (RS1-RS39)
1299.000-1299.750	150kHz	DD High Speed Digital Data – 5 x 150kHz channels
All Modes		1299.075, 1299.225, 1299.375, 1299.525, 1299.675MHz (±75kHz)

1.3GHz (23cm)

IARU Recommendation	NECESSARY BANDWIDTH	UK USAGE	Contd
1299.750-1300.000 All Modes	20kHz	25kHz Channels Available for FM/DV use 1299.775-1299.975MHz	
1300.000-1325.000 ATV		TV Repeaters (UK only) (Note 9) New DATV Repeater Outputs Original ATV Repeater Outputs: 1308.0, 1310.0, 1311.5, 1312.0, 1316.0, 1318.5MHz	

Note 1: Local traffic using narrowband modes should operate between 1296.500-1296.800MHz during contests and band openings.
Note 2: Stations in countries that do not have access to 1298-1300MHz may also use the FM simplex segment for digital communications.
Note 3: IARU Region 1 recommended maximum bandwidth is 20kHz. See also Note 7.
Note 4: deleted.
Note 5: Embedded data traffic is allowed with digital voice (DV).
Note 6: Simplex use only – no DV gateways.
Note 7: 1240.000-1240.750 has been designated by IARU as an alternative centre for narrowband activity and beacons. Operations in this range should be on a flexible basis to enable coordinated activation of this alternate usage.
Note 8: The band 1240-1300MHz is subject to major replanning. Contact the Microwave Manager for further information.
Note 9: Repeaters and Migration to DATV, inc option for new DATV simplex are subject to further development and coordination.
Note 10: QPSK 4 Mega-symbols/second maximum recommended.
Licence Notes: Amateur Service – Secondary User. Amateur Satellite Service: 1,260-1,270MHz – Secondary User Earth to Space only. In the sub-band 1,298-1,300MHz unattended operation is not allowed within 50km of SS206127 (Bude), SE202577 (Harrogate), or in Northern Ireland.
Notes to the Band Plan: As on page 167.

2.3-2.302GHz

IARU Recommendation	NECESSARY BANDWIDTH	UK USAGE

Access to this band requires an appropriate NoV, which is available to Full licensees only. Please note that the current NoVs last for up to three years prior to expiry.

2300.000-2300.400MHz	2.7kHz	Narrowband Modes (including CW, SSB, MGM) 2300.350-2300.400MHz Attended Beacons
2300.400-2301.800MHz	500kHz	Wideband Modes (NBFM, DV, Data, DATV, etc) Note 1
2301.800-2302.000MHz	2.7kHz	Narrowband modes (including CW, SSB, MGM) EME Usage

Note 1: Users of wideband modes must ensure their spectral emissions are contained within the band limits.
Note 2: Full licensees only with NoV, 400 watts maximum, not available in the Isle of Man or Channel isles. Note additional restrictions on usage are specified by the NoV terms. It should be emphasised that this is UK-specific and is available on a non interference basis to exisiting services.
Notes to the Band Plan: As on page 167.

2.3GHz (13cm)

IARU Recommendation	NECESSARY BANDWIDTH	UK USAGE
2,310.000-2,320.000MHz (National band plans)	200kHz	2,310.000-2,310.500MHz – Repeater links
		2,311.000-2,315.000MHz – High speed data Preferred Narrowband Segment
2,320.000-2,320.150	500Hz	2,320.000-2,320.025MHz – Moonbounce
2,320.150-2,320.800	2.7kHz	2,320.200MHz – SSB Centre of Activity
		2,320.750-2,320.800MHz – Local Beacons, 10W ERP max
2,320.800-2,321.000		2,320.800-2,320.990MHz – Propagation Beacons Only
Beacons exclusive 2321.000-2322.000	20kHz	FM/DV. See also Note 1
2,322.000-2,350.000		Wideband Modes including Data, ATV
2,390.000-2,400.000		All Modes
2,400.000-2,450.000MHz Satellites		2,435.000MHz ATV Repeater Outputs 2,440.000MHz ATV Repeater Outputs

Note 1: Stations in countries which do not have access to the All Modes section 2,322-2,390MHz, use the simplex and repeater segment 2,320-2,322MHz for data transmission.
Note 2: Stations in countries that do not have access to the narrowband segment 2,320-2,322MHz, use the alternative narrowband segment 2,304-2,306MHz and 2,308-2,310MHz.
Note 3: The segment 2,433-2,443MHz may be used for ATV if no satellite is using the segment.
Licence Notes: Amateur Service – Secondary User. Users must accept interference from ISM users. Amateur Satellite Service: 2,400-2,450MHz – Secondary User. Users must accept interference from ISM users. Operation in 2310-2350 and 2390-2400 MHz are subject to specific conditions and guidance In the sub-bands 2,310.000-2,310.4125 and 2,392-2,450MHz unattended operation is not allowed within 50km of SS206127 (Bude) or SE202577 (Harrogate). ISM = Industrial, scientific and medical.
Notes to the Band Plan: As on page 167.

3.4GHz (9cm)

IARU Recommendation	NECESSARY BANDWIDTH	UK USAGE
3,400.000-3,401.000MHz	2.7kHz	Narrowband Modes (including CW, SSB, MGM, EME) 3,400.100MHz – Centre of Activity (Note 1)
		3,400.750-3,400.800MHz – Local Beacons, 10W ERP max
3,400.800-3,400.995		3,400.800-3,400.995MHz – Propagation Beacons Only
Propagation Beacons 3,400.000-3,401.000MHz 3,402.000-3,410.000 Outputs All Modes (Notes 2, 3)	200kHz	3,401.000-3,402.000MHz Data, Remote Control Wideband Modes including DATV Repeater

Note 1: EME has migrated from 3456MHz to 3400MHz to promote harmonised usage and activity.
Note 2: Stations in many European countries have access to 3400-3410MHz as permitted by ECA Table Footnote EU17.
Note 3: Amateur Satellite downlinks planned.
Licence Notes: Amateur Service – Secondary User. Subject to specific conditions and guidance.
Notes to the Band Plan: As on page 167.

5.7GHz (6cm)

IARU Recommendation	UK USAGE
5,650.000-5,668.000MHz Satellite Uplinks	Amateur Satellite Service – Earth to Space Only
5,650.000-5,670.000 Narrowband CW/EME/SSB	5,668.200MHz – Alternative Centre of Activity 5,668.8MHz – Beacons
5,670.000-5,680.000 All Modes	
5,755.000-5,760.000 All Modes	
5,760.000-5,762.000 Narrowband CW/EME/SSB	5,760.100MHz – Current Centre of Activity
	5,760.750-5,760.800MHz – Local Beacons, 10W ERP max
5760.800-5760.995	5,760.800-5,760.995MHz – Propagation Beacons only
Propagation Beacons 5,762.000-5,765.000 All Modes	
5,820.000-5,830.000 All Modes	
5,830.000-5,850.000 Satellite Downlinks	Amateur Satellite Service – Space to Earth Only

Licence Notes: Amateur Service: 5,650-5,680MHz – Secondary User. 5,755-5,765 and 5,820-5,850MHz – Secondary User. Users must accept interference from ISM users. Amateur Satellite Service: 5,650-5,670MHz and 5,830-5,850MHz – Secondary User. Users must accept interference from ISM users. Unattended operation is permitted for remote control, digital modes and beacons, except in the sub-bands 5,670-5,680MHz within 50km of SS206127 (Bude) and SE202577 (Harrogate). ISM = Industrial, scientific and medical.
Notes to the Band Plan: As on page 167.

10GHz (3cm)

IARU Recommendation	NECESSARY BANDWIDTH	UK USAGE
10,000.000-10,125.000MHz All Modes		Note 4 10,065MHz ATV Repeater Outputs
10,225.000-10,250.000 All Modes		10,240MHz ATV Repeaters
10,250.000-10,350.000 Digital Modes		
10,350.000-10,368.000		10,352.5-10,368MHz Wideband Modes (Note 2)
All Modes 10,368-10,370MHz Narrowband Telegraphy EME/SSB	2.7kHz	10,368-10,370 Narrowband Modes (Note 3) 10,368.1MHz Centre of Activity
		10,368.750-10,368.800MHz – Local Beacons, 10W ERP max
10,368.800-10,368.995		10,368.800-10,368.995MHz – Propagation Beacons Only
Propagation Beacons 10,370.000-10,450.000 All Modes		10,371MHz Voice Repeaters Rx
10,450.000-10,475.000 All Modes & Satellites		10,425 ATV Repeaters 10,400-10,475MHz Unattended Operation 10,450-10,452MHz Alternative Narrowband Segment (Note 3) 10,471MHz Voice Repeaters Tx
10,475.000-10,500.000 All Modes and satellites		Amateur Satellite Service ONLY

Note 1: Deleted.
Note 2: Wideband FM is preferred between 10,350-10,400MHz to encourage compatibility between narrowband systems.
Note 3: 10,450MHz is used as an alternative narrowband segment in countires where 10,368MHz is not available.
Note 4: 10,000-10,125MHz is subject to increased Primary user utilisation and NoV restrictions.
Note 5: 10,475-10,500MHz is allocated ONLY to the Amateur Satellite Service and NOT to the Amateur Service.
Licence Notes: Amateur Service – Secondary User. Foundation licensees 1 watt maximum. Amateur Satellite Service: 10,450-10,500MHz – Secondary User. Unattended operation is permitted for remote control, digital modes and beacons except in the sub-bands 10,000-10,125MHz within 50km of SO916223 (Cheltenham), SS206127 (Bude), SK985640 (Waddington) and SE202577 (Harrogate).
Notes to the Band Plan: As on page 167.

24GHz (12mm) IARU Recommendation	UK USAGE
24,000.000-24,050.000MHz Satellites Equipment	24,025MHz Preferred Operating Frequency for Wideband
	24,048.2MHz – Narrowband Centre of Activity **24,048.750-24,048.800MHz – Local Beacons, 10W ERP max**
Propagation Beacons 24,050.000-24,250.000 All Modes	

Licence Notes: Amateur Service: 24,000-24,050MHz – Primary User: Users must accept interference from ISM users. 24,050-24,150MHz – Secondary User. May only be used with the written permission of Ofcom. Users must accept interference from ISM users. 24,150-24,250MHz – Secondary User. Users must accept interference from ISM users. Amateur Satellite Service: 24,000-24,050MHz – Primary User: Users must accept interference from ISM users. Unattended operation is permitted for remote control, digital modes and beacons, except in the sub-bands 24,000-24,050MHz within 50km of SK985640 (Waddington) and SE202577 (Harrogate).
ISM = Industrial, scientific and medical.
Notes to the Band Plan: As on page 167.

47GHz (6mm) IARU Recommendation	UK USAGE
47,000.000-47,200.000MHz 47,088.000-47,090.000 Narrowband Segment	47,088.2MHz – Centre of Narrowband Activity **47,088.8-47,089.0MHz – Propagation Beacons Only**

Licence Notes: Amateur Service and Amateur Satellite Service – Primary User. Unattended operation is permitted for remote control, digital modes and beacons, except within 50km of SK985640 (Waddington) and SE202577 (Harrogate).
Notes to the Band Plan: As on page 167.

76GHz (4mm) IARU Recommendation	UK USAGE
75,500-76,000MHz All Modes (preferred)	75,976.200MHz – IARU Region 1 Preferred Centre of Activity
76,000.000-77,500.000 All Modes	
77,500-78,000 Segment All Modes (preferred)	77,500.200MHz – Alternative IARU Recommended Narrowband
78,000-81,000 All Modes	

Licence Notes:
75,500-75,875MHz Amateur Service and Amateur Satellite Service – Secondary User.
75,875-76,000MHz Amateur Service and Amateur Satellite Service – Primary User.
76,000-77,500MHz Amateur Service and Amateur Satellite Service – Secondary User.
77,500-78,000MHz Amateur Service and Amateur Satellite Service – Primary User.
78,000-81,000MHz Amateur service and Amateur Satellite Service – Secondary User.
Unattended operation is permitted for remote control, digital modes and beacons, except within 50km of SK985640 (Waddington) and SE202577 (Harrogate).
Notes to the Band Plan: As on page 167.

134GHz (2mm) IARU Recommendation	UK USAGE
134,000-134,928MHz All Modes	
134,928 -134,930 Narrowband Modes	IARU Region 1 Preferred Centre of Activity
134,930 -136,000 All Modes	**134,928.800-134,928.990 – Propagation Beacons Only**

Licence Notes: 134,000-136,000MHz Amateur Service and Amateur Satellite Service – Primary User. Unattended operation is permitted for remote control, digital modes and beacons, except within 50km of SK985640 (Waddington) and SE202577 (Harrogate).

THE FOLLOWING BANDS ARE ALSO ALLOCATED TO THE AMATEUR SERVICE AND THE AMATEUR SATELLITE SERVICE

122,250-123,000MHz – Amateur Service only, Secondary User
136,000-141,000MHz – Secondary User
241,000-248,000MHz – Secondary User
248,000-250,000MHz – Primary User
Notes to the Band Plan: As on page 167.

NOTES TO THE BAND PLAN

ITU-R Recommendation SM.328 (extract)

Necessary bandwidth: For a given class of emission, the width of the frequency band which is just sufficient to ensure the transmission of information at the rate and with the quality required under specified conditions.

Foundation and Intermediate Licence holders are advised to check their Licences for the permitted power limits and conditions applicable to their class of Licence.

All Modes: CW, SSB and those modes listed as Centres of Activity, plus AM. Consideration should be given to adjacent channel users.

Image Modes: Any analogue or digital image modes within the appropriate bandwidth, for example SSTV and FAX.

Narrowband Modes: All modes using up to 500Hz bandwidth, including CW, RTTY, PSK, ie.modes.

Digimodes: Any digital mode used within the appropriate bandwidth, for example RTTY, PSK, MT63, etc.

Sideband usage: Below 10MHz use lower sideband (LSB), above 10MHz use upper sideband (USB). Note the lowest dial settings for LSB Voice modes are 1843, 3603 and 7043kHz on 160, 80 and 40m. Note that on (5MHz) USB is used.

Amplitude Modulation (AM): AM with a bandwidth greater than 2.7kHz is acceptable in the All Modes segments provided users consider adjacent channel activity when selecting operating frequencies (Davos 2005).

Extended SSB (eSSB): Extended SSB (eSSB) is only acceptable in the All Modes segments provided users consider adjacent channel activity when selecting operating frequencies.

Digital Voice (DV): Users of Digital Voice (DV) should check that the channel is not in use by other modes (CT08_C5_Rec20).

FM Repeater & Gateway Access: CTCSS Access is recommended. Toneburst access is being withdrawn in line with IARU-R1 recommendations.

Beacons: Propagation Beacon Sub-bands are highlighted – please avoid transmitting in them!

MGM: Machine Generated Modes indicates those transmission modes relying fully on computer processing such as RTTY, AMTOR, PSK31, JTxx, FSK441 and the like. This does not include Digital Voice (DV) or Digital Data (DD).

WSPR: Above 30MHz, WSPR frequencies in the band plan are the centre of the transmitted frequency (not the suppressed carrier frequency or the VFO dial setting).

CW QSOs are accepted across all bands, except within beacon segments (Recommendation DV05_C4_Rec_13).

Contest activity shall not take place on the 10, 18 and 24MHz (30, 17 and 12m) bands.

Non-contesting radio amateurs are recommended to use the contest-free HF bands (30, 17 and 12m) during the largest international contests (DV05_C4_Rev_07).

The term 'automatically controlled data stations' include Store and Forward stations.

Transmitting Frequencies: The announced frequencies in the band plan are understood as 'transmitted frequencies' (not those of the suppressed carrier!).

Unmanned transmitting stations: IARU member societies are requested to limit this activity on the HF bands. It is recommended that any unmanned transmitting stations on HF shall only be activated under operator control except for beacons agreed with the IARU Region 1 Beacon Coordinator, or specially licensed experimental stations.

472-479kHz: Access is available to Full licensees only - see licence schedule for additional conditions.

1.8MHz: Radio amateurs in countries that have a SSB allocation ONLY below 1840kHz, may continue to use it, but the National Societies in those countries are requested to take all necessary steps with their licence administrations to adjust phone allocations in accordance with the Region 1 Band Plan (UBA – Davos 2005).

3.5MHz: Inter-Continental operations should be given priority in the segments 3500-3510kHz and 3775- 3800kHz. Where no DX traffic is involved, the contest segments should not include 3500-3510kHz or 3775-3800kHz. Member societies will be permitted to set other (lower) limits for national contests (within these limits). 3510-3600kHz may be used for unmanned ARDF beacons (CW, A1A) (Recommendation DV05_C4_Rec_12).

Member societies should approach their national telecommunication authorities and ask them not to allocate frequencies other than amateur stations in the band segment that IARU has assigned to Inter-Continental long distance traffic.

5MHz: Access is available to Full licensees only- see licence schedule for additional conditions.

7MHz: The band segment 7040-7060kHz may be used for automatic controlled data stations (unattended) traffic in the areas of Africa south from the equator during local daylight hours. Where no DX traffic is involved, the contest segment should not include 7,175-7,200kHz.

10MHz: SSB may be used during emergencies involving the immediate safety of life and property and only by stations actually involved in the handling of emergency traffic.

The band segment 10120kHz to 10140kHz may be used for SSB transmissions in the area of Africa south of the equator during local daylight hours.

News bulletins on any mode should not be transmitted on the 10MHz band.

28MHz: Member societies should advise operators not to transmit on frequencies between 29.3 and 29.51MHz to avoid interference to amateur satellite downlinks.

Experimentation with NBFM Packet Radio on 29MHz band: Preferred operating frequencies on each 10kHz from 29.210 to 29.290MHz inclusive should be used. A deviation of ±2.5kHz being used with 2.5kHz as maximum modulation frequency.

146-147MHz & 2300-2302MHz
Access to these bands requires an appropriate NoV, which is available to Full licensees only.

430MHz
The use of Amplitude Modulation (AM) is acceptable in the all modes segments but users are asked to consider

1.3GHz
The band is subject to re-planning. It is also shared with air traffic radar.

2.3GHz (2310-2350 & 2390-2400MHz)
Operation is subject to specific licence conditions and guidance - see also the Ofcom PSSR statement.

3.4GHz (3400-3410MHz)
Operation is subject to specific licence conditions and guidance - see also the Ofcom PSSR statement.

Repeater Listings

There are over 700 repeaters licensed in the United Kingdom. They range in frequency from 28MHz to 10GHz. Many are traditional FM units, but there are also several for amateur television. Repeaters are increasingly being connected to the Internet and D-STAR, FUSION and other digital repeaters are increasing in number.

Repeaters (alphabetical)

Callsign	Channel	Locator	Location
GB3AA	RV53	IO81RO	Bristol
GB3AB	RB07	IO93FK	Sheffield
GB3AC	RU78	IO81SR	Lydney Glos
GB3AG	RV58	IO86ON	Forfar
GB3AH	RB11	JO02KP	East Dereham
GB3AK	RM14A	IO81RO	Bristol
GB3AL	RV59	IO91QP	Amersham
GB3AM	R50-13	IO91QP	Amersham
GB3AN	RB08	IO73UJ	Amlwch
GB3AR	RV56	IO73VC	Caernarfon
GB3AS	RV48	IO85MC	Langholm
GB3AU	RB07	IO91QP	Amersham
GB3AV	RB02	IO91OT	Aylesbury
GB3AW	RB10	IO91HH	Newbury
GB3AY	RV52	IO75OR	Dalry
GB3BB	RV56	IO81HW	Brecon
GB3BE	RU69	IO85SS	Duns
GB3BF	RV63	IO92SD	Bedford
GB3BG	RB09	IO81KS	Blaenavon
GB3BI	RV58	IO77WO	Inverness
GB3BK	RM0A	JO01AK	Bromley
GB3BL	RB07	IO92SD	Bedford
GB3BM	RV57	IO92BL	Birmingham
GB3BN	RB0	IO91OJ	Bracknell
GB3BP	RU71	IO93GA	Belper
GB3BR	RB06	IO90WT	Brighton
GB3BS	RU68	IO81TK	Bristol
GB3BT	RV56	IO85WT	Berwick On Tweed
GB3BV	RB01	IO91SR	Hemel Hempstead
GB3BX	RV54	IO82QN	Much Wenlock
GB3BZ	RU68	JO01GW	Braintree
GB3CA	RB13	IO84OT	Carlisle
GB3CB	RB14	IO92BL	Birmingham
GB3CC	RU77	IO90RM	Chichester
GB3CD	RV55	IO94DR	Crook
GB3CE	RB14	JO01KV	Colchester
GB3CF	RV48	IO92IQ	Markfield
GB3CG	RV58	IO81VU	Gloucester
GB3CH	RB02	IO70SM	Liskeard
GB3CI	RU66	IO92PM	Corby
GB3CJ	10M	IO92NF	Northampton
GB3CK	RB0	JO01JF	Charing Kent
GB3CL	RB09	JO01OT	Clacton
GB3CM	RB08	IO71VW	Carmarthen
GB3CO	RV53	IO92PM	Corby
GB3CP	RV59	IO64IG	Fermanagh
GB3CR	RB06	IO83LC	Caergwrle
GB3CS	RV60	IO85AU	Motherwell
GB3CV	RB09	IO92GK	Coventry
GB3CW	RB04	IO82HL	Newtown Powys
GB3DA	RV58	JO01GR	Danbury
GB3DB	R50-06	JO01HR	Danbury
GB3DC	RV55	IO92GW	Derby
GB3DD	RB10	IO86MM	Dundee
GB3DE	RB07	JO02NF	Ipswich
GB3DG	RV62	IO74UV	Newton Stewart
GB3DI	RB06	IO91IN	Didcot
GB3DM	RU74	IO75RX	Dumbarton
GB3DN	RV51	IO70UW	Stibb Cross
GB3DQ	RU78	IO70RI	Polperro
GB3DR	RV59	IO80QR	Dorchester
GB3DS	RB13	IO93KH	Worksop
GB3DT	RB0	IO80WU	Blandford
GB3DU	RV61	IO85SS	Duns
GB3DV	RB01	IO93JK	Maltby
GB3DW	RV53	IO72WT	Harlech
GB3DX	RB12	IO65HA	Derry/Londonderry
GB3DY	RB10	IO93FB	Wirksworth
GB3EA	RV55	JO02GE	Wickhambrook
GB3EB	RV63	JO00BW	Uckfield
GB3EC	R50-12	JO02HE	Bury St Edmunds
GB3ED	RB14	IO85JW	Edinburgh
GB3EE	RB12	IO93FE	Chesterfield
GB3EF	R50-01	JO02NF	Stowmarket
GB3EH	RB08	IO92GC	Edge Hill
GB3EI	RV49	IO67IN	Clachan North Uist
GB3EK	RU71	JO01QJ	Margate
GB3EL	RV48	JO01AM	London
GB3ER	RB03	JO01GR	Danbury
GB3ES	RV54	JO00HV	Hastings
GB3EV	RV56	IO84SQ	Dufton
GB3EW	RV49	IO80FR	Exeter
GB3EX	RU76	IO80GT	Silverton
GB3EZ	RU78	JO02GE	Wickhambrook
GB3FC	RU68	IO83LU	Blackpool
GB3FE	RV53	IO86BC	Stirling
GB3FF	RV48	IO86GC	Kelty
GB3FG	RV57	IO71UX	Carmarthen
GB3FH	R50-06	IO81OH	Somerset
GB3FI	RU74	IO81OH	Cheddar
GB3FJ	RU76	IO83XE	Asgarby
GB3FK	RV60	JO01OC	Folkestone
GB3FM	RM2	IO91OF	Farnham
GB3FN	RB15	IO91OF	Farnham
GB3FR	RV62	JO03AE	Spilsby Lincs.
GB3FX	R50-10	IO91OF	Farnham
GB3GB	RB12	IO92BN	Birmingham
GB3GC	R50-02	IO70WN	Gunnislake
GB3GD	RV50	IO74SG	Snaefell IoM
GB3GF	RB12	IO91RF	Guildford
GB3GH	RB05	IO81VU	Gloucester
GB3GJ	RV51	IN89WE	St Helier
GB3GL	RU72	IO75WU	Glasgow
GB3GN	RV62	IO87TA	Banchory
GB3GO	RV63	IO83BH	Llandudno
GB3GR	RU68	IO92QV	Grantham
GB3GT	R50-12	IO82QJ	Ludlow
GB3GU	RB13	IN89RK	Guernsey
GB3GX	10M	IO82QJ	Ludlow
GB3GY	RB11	IO93XN	Cleethorpes
GB3HA	RV60	IO85XA	Corbridge
GB3HB	RB15	IO70OJ	Roche
GB3HC	RB06	IO82PB	Hereford
GB3HD	RB09	IO93BP	Huddersfield
GB3HE	RB14	JO00HV	Hastings
GB3HF	R50-05	IO93HO	Barnsley
GB3HG	RV50	IO94JF	Thirsk
GB3HH	RV56	IO93BF	Buxton
GB3HI	RV56	IO76DK	Isle Of Mull
GB3HJ	RB01	IO94EB	Harrogate
GB3HK	RB02	IO85ON	Selkirk
GB3HM	R50-11	IO93GA	Belper
GB3HO	RU71	IO91TB	Horsham
GB3HR	RB14	IO91TO	Harrow
GB3HS	RV52	IO93RT	Walkington
GB3HT	RB11	IO91HN	Hinckley
GB3HW	RB13	JO01CN	Gidea Park
GB3HY	RU72	IO90WX	Haywards Heath
GB3IC	RU65	IO82WO	Wolverhampton
GB3IE	RU68	IO70XJ	Plymouth
GB3IG	RV62	IO68QE	Stornoway
GB3IH	RB04	JO02OB	Ipswich
GB3IK	RV61	JO01FJ	Rochester
GB3IM	RU66	IO74SD	Douglas IoM
GB3IM	RU70	IO74PF	Peel
GB3IM	RU66	IO74TI	Ramsey IoM
GB3IM	RB05	IO74SG	Douglas IoM
GB3IN	RV51	IO93GD	Alfreton
GB3IP	RV61	IO92WS	Stafford
GB3IR	RV61	IO94DJ	Richmond Yorks
GB3IS	RV60	IO77BF	Broadford
GB3IW	RB09	IO90JQ	Ryde
GB3JB	RV63	IO81VC	Mere Wiltshire
GB3JC	RU72	JO02QP	Norwich
GB3JL	RV63	IO70SM	Liskeard
GB3JS	RB01	JO02UO	Great Yarmouth
GB3JU	RU72	IN89WE	St Helier
GB3KC	RU72	IO82WK	Stourbridge
GB3KD	RV63	IO82UJ	Kidderminster
GB3KE	RV55	IO75UV	Glasgow
GB3KI	RV53	JO01NI	Herne Bay
GB3KK	RU78	IO65VE	Ballycastle
GB3KL	RB04	JO02FR	Kings Lynn
GB3KN	RV56	JO01HH	Maidstone
GB3KR	RB03	IO82UJ	Kidderminster
GB3KS	RV50	JO01PA	Dover
GB3KU	RB03	IO83XM	Ashton-U-Lyne
GB3KV	RU78	IO75VX	Kilsyth
GB3KW	RU66	IO75UV	Glasgow
GB3KY	RV57	JO02FS	Kings Lynn
GB3LA	RV57	IO85CJ	Sanquhar
GB3LB	RV63	IO85NS	Lauder
GB3LC	RB09	IO93WH	Louth
GB3LD	RV54	IO84OA	Lancaster
GB3LE	RB04	IO92IQ	Markfield
GB3LF	RB14	IO84PH	Kendal
GB3LH	RB15	IO82OP	Shrewsbury
GB3LI	RB10	IO83LL	Liverpool
GB3LL	RB0	IO83BH	Llandudno
GB3LM	RV58	IO93RF	Lincoln
GB3LP	R50-04	IO83MK	Liverpool
GB3LR	RU69	JO00AS	Newhaven
GB3LS	RB02	IO93RF	Lincoln
GB3LT	RB10	IO91SV	Luton
GB3LU	RV54	IP90JD	Lerwick
GB3LV	RB02	IO91XP	North London
GB3LW	RU72	IO91WM	London
GB3LY	RV48	IO65NC	Limavady
GB3MA	RB01	IO83UO	Bury
GB3MB	RB15	IO81HS	Merthyr Tydfil
GB3ME	RB06	IO92JJ	Rugby
GB3MF	RB02	IO83WG	Macclesfield
GB3MH	RV50	IO91WC	East Grinstead
GB3MI	RV57	IO83VM	Manchester
GB3MK	RB0	IO92OB	Milton Keynes
GB3MM	RM6	IO82XP	Wolverhampton
GB3MN	RV52	IO83XH	Disley
GB3MP	RV60	IO83IF	Denbigh
GB3MR	RB14	IO83XH	Stockport
GB3MS	RB07	IO82VE	Worcester
GB3MT	RU72	IO64QS	Magherafelt
GB3MW	RB10	IO92FH	Leamington Spa
GB3NA	RV49	IO93HO	Barnsley
GB3NB	RV50	JO02PN	Norwich
GB3NC	RV58	IO70OJ	Roche
GB3ND	RB14	IO71VA	Bideford
GB3NE	RV61	IO91HJ	Newbury
GB3NF	RV50	IO92KX	Nottingham
GB3NG	RV50	IO87XO	Fraserburgh
GB3NH	RU70	IO92NF	Northampton
GB3NI	RV58	IO74CO	Belfast N.I.
GB3NK	RB04	JO01BL	Erith
GB3NL	RV62	IO91XP	North London
GB3NM	RB07	IO92KX	Nottingham
GB3NN	RB02	JO02JV	Wells Norfolk
GB3NO	RM0	JO02PP	Norwich
GB3NP	RU71	IO92LD	Towcester
GB3NR	RB0	JO02PP	Norwich
GB3NS	DVU54	IO91WG	Caterham
GB3NT	RB0	IO94FW	Newcastle Upon Tyne
GB3NU	RU71	JO02OW	Sheringham
GB3NW	RV50	IO82VE	Worcester
GB3NX	RU68	IO91WB	Crawley Sussex
GB3NY	RU66	IO94VC	Bridlington
GB3NZ	RV48	JO02QP	Thorpe St Andrew
GB3OA	RV49	IO83LP	Southport
GB3OC	RV52	IO88LX	Kirkwall
GB3OH	RU76	IO85EX	Linlithgow
GB3OM	RU76	IO64JQ	Omagh
GB3OV	RB05	IO92VG	St Neots
GB3OY	RU76	JO01AO	Buckhurst Hill
GB3PA	RV50	IO75QV	Paisley
GB3PE	RV54	IO92TN	Peterborough
GB3PI	RV60	IO92XA	Royston
GB3PK	RV53	IO65WE	Ballycastle N.I.
GB3PL	RV61	IO70VM	E.Cornwall
GB3PO	RV52	JO02OB	Ipswich
GB3PP	RB15	IO83PS	Preston
GB3PR	RV54	IO86GI	Perth
GB3PS	RM3	IO92XA	Royston
GB3PU	RB0	IO86GI	Perth
GB3PW	RV62	IO82HL	Newtown Powys
GB3PX	R50-07	IO92XA	Royston
GB3PY	RB08	JO02AF	Cambridge
GB3PZ	RU72	IO83XL	Dukinfield
GB3RA	RV51	IO82JH	Llandrindod Wells
GB3RB	RB08	IO93IF	Bolsover
GB3RD	RV54	IO91JM	Reading
GB3RE	RB11	JO01HH	Maidstone
GB3RF	RV62	IO83TR	Accrington

Repeaters (alphabetical)

Callsign	Channel	Locator	Location
GB3RH	RB11	IO80MS	Axminster
GB3RT	RB0	IO81LP	Cwmbran
GB3RU	RB11	IO91LK	Reading
GB3SB	RV52	IO85ON	Selkirk
GB3SD	RB14	IO80SQ	Weymouth
GB3SE	RM3	IO83WA	Stoke On Trent
GB3SF	RU75	IO81DA	South Molton
GB3SH	RV53	IO90GX	Romsey
GB3SI	RV50	IO70JB	Helston
GB3SK	RB06	JO01MH	Canterbury
GB3SL	R50-02	IO75XX	Kilsyth
GB3SM	RB13	IO93BA	Stoke On Trent
GB3SN	RV58	IO91LC	Alton Hants
GB3SP	RB04	IO71NQ	Pembroke
GB3SS	RV48	IO87MO	Elgin Moray
GB3ST	RB09	IO83WA	Stoke On Trent
GB3SW	RV57	IO80JR	Sidmouth
GB3SX	R50-08	IO93BA	Stoke On Trent
GB3SY	RB09	IO93HO	Barnsley
GB3TA	RV59	IO92DP	Tamworth
GB3TD	RB03	IO91DL	Swindon
GB3TE	RV49	JO01MT	Clacton On Sea
GB3TF	RB08	IO82RP	Telford
GB3TH	RB15	IO92DP	Tamworth
GB3TJ	RB12	IO85XA	Corbridge
GB3TO	RV49	IO92NE	Northampton
GB3TP	RV58	IO93CT	Shipley
GB3TR	RV52	IO80FM	Torquay
GB3TS	RB07	IO94GX	Sunderland
GB3TU	RB09	IO91PS	Tring Herts
GB3TW	RV58	IO94EW	Gateshead
GB3TY	R50-07	IO74BS	Carrickfergus
GB3UB	RB04	IO81UJ	Bath Avon
GB3UK	RU69	IO81XW	Cheltenham
GB3UK	RU69	IO81XW	Cheltenham
GB3UL	RB02	IO74CO	Belfast N.I.
GB3UM	R50-03	IO92IQ	Markfield
GB3UO	RU66	IO82LW	Chirk
GB3US	RB0	IO93GI	Sheffield
GB3VA	RV56	IO91LT	Brill
GB3VE	RB04	IO84SQ	Dufton
GB3VH	RB13	IO91VT	Welwyn Garden City
GB3VI	R50-15	IO92BM	Birmingham
GB3VM	RV49	IO82PH	Ludlow
GB3VN	RU74	IO82QI	Ludlow
GB3VO	RV51	IO82LW	Chirk
GB3VS	RB03	IO80LX	Taunton
GB3VT	RV57	IO83WA	Stoke On Trent
GB3WA	RU70	IO75OR	Dalry
GB3WB	RB12	IO81MH	Weston-S-Mare
GB3WC	RM15	IO93EP	Wakefield
GB3WD	RV56	IO70WJ	Plymouth
GB3WE	RV55	IO81MH	Weston-S-Mare
GB3WF	RB14	IO93DV	Otley
GB3WG	RU70	IO81CP	Port Talbot
GB3WH	RV52	IO91EM	Swindon
GB3WI	RB15	JO02BP	Wisbech
GB3WJ	RB05	IO93QN	Scunthorpe
GB3WK	RV62	IO92FH	Leamington Spa
GB3WL	RU76	IO92BJ	Birmingham
GB3WN	RB0	IO82XP	Wolverhampton
GB3WO	RU78	IO90UT	Lancing
GB3WP	RU75	IO83XL	Hyde
GB3WR	RV48	IO81PH	Cheddar
GB3WS	RV60	IO91WB	Crawley Sussex
GB3WT	RV62	IO64JQ	Omagh
GB3WU	RU66	IO82VE	Worcester
GB3WW	RV62	IO81CP	Port Talbot
GB3WX	10M	IO81VC	Mere Wiltshire
GB3WX	6M	IO81VC	Mere Wiltshire
GB3WY	R50-09	IO93EP	Wakefield
GB3XD	R50-02	IO93WH	Louth
GB3XL	RU71	IO93CT	Shipley
GB3XN	RU74	IO93KJ	Worksop
GB3XP	RV55	IO91VJ	Morden
GB3XX	RV73	IO92KG	Daventry
GB3YC	RV60	IO94SC	Driffield
GB3YL	RB14	JO02VL	Lowestoft
GB3YS	RB02	IO80QW	Yeovil
GB3YW	RV63	IO93EP	Wakefield
GB3ZA	RV59	IO82PB	Hereford
GB3ZB	RU66	IO81QJ	Bristol
GB3ZI	RU78	IO82WT	Stafford
GB3ZW	R50-10	IO82HL	Newtown Powys
GB3ZX	RU66	JO00CT	Eastbourne
GB3ZY	R50-09	IO81QJ	Bristol
GB7AA	DVU54	IO81RO	Bristol
GB7AD	DVU61	IO81RO	Bristol
GB7AH	DVU41	IO64TU	Ahoghill
GB7AK	DVU42	JO00AX	Barking
GB7AL	DVU32	JO02RD	Ipswich
GB7AS	DVU34	JO01KD	Ashford
GB7AU	DVU56	IO91QP	Amersham
GB7AV	DVU35	IO91OT	Aylesbury
GB7BD	RU76	IO81QJ	Bristol
GB7BE	DVU53	JO02TK	Beccles
GB7BJ	DVU59	IO82QE	Bromyard
GB7BK	DVU59	IO91JM	Reading
GB7BM	DVU32	IO92BL	Birmingham
GB7BP	DVU36	IO91PX	Milton Keynes
GB7BR	RU74	IO74TI	Ramsey IoM
GB7BS	DVU13	IO81TK	Bristol
GB7BX	DVU35	IO82QN	Much Wenlock
GB7CA	RU74	IO74SD	Douglas
GB7CD	DVU56	IO81JL	Cardiff
GB7CF	DVU41	JO01OC	Folkestone
GB7CH	DVU38	IO93WH	Louth
GB7CK	DVU59	JO01OC	Folkestone
GB7CL	DVU51	JO01MT	Clacton On Sea
GB7CM	RV56	IO80WU	Blandford
GB7CS	DVU62	IO75VS	East Kilbride
GB7CT	RV51	IO81EN	Tring
GB7CW	DVU32	IO81EN	Bridgend
GB7DA	RV62	IO85AV	Airdrie
GB7DB	DVU53	IO92RA	Ampthill
GB7DC	DVU39	IO92GW	Derby
GB7DD	DVU53	IO86NL	Dundee
GB7DE	RV51	IO85XA	Edinburgh
GB7DG	DVU43	IO82WI	Bromsgrove
GB7DJ	DVU53	IO83RF	Northwich
GB7DK	RV55	IO74LV	Stranraer
GB7DK-B	DVU52	IO74LV	Stranraer
GB7DL	DVU54	JO03BB	Friskney
GB7DN	DVU33	IO64MW	Dungiven
GB7DO	RU69	IO93JO	Doncaster
GB7DP-B	DVU40	IO86LL	Dundee
GB7DR	DVU34	IO90AR	Poole
GB7DS	DVU34	JO02PP	Norwich
GB7DV	DVU49	IO83QJ	St. Helens
GB7DX	DVU56	JO01KD	Ashford Kent
GB7EB	DVU46	JO02TK	Beccles
GB7ED	DVU54	IO80GR	Exeter
GB7EE	DVU57	IO85JW	Edinburgh
GB7EG	DVU61	IO91XC	East Grinstead
GB7EK	DVU50	JO01NI	Whitstable
GB7EL	DVU57	IO83VT	Nelson
GB7EN	DVU43	IO92PH	Wellingborough
GB7EP	DVU32	IO91OT	Epsom
GB7ER	RV62	IO80GR	Exeter
GB7ES	DVU35	JO00DT	Eastbourne
GB7EX	DVU57	JO01GN	Southend On Sea
GB7FC	RU78	IO83LU	Blackpool
GB7FF	DVU58	IO83LT	Blackpool
GB7FG	RU76	IO71WW	Carmarthen
GB7FH	DVU49	IO90JU	Fareham
GB7FI	RU71	IO81OH	Axbridge
GB7FK-C	RV63	JO01OC	Folkestone
GB7FO	DVU35	IO83LT	Blackpool
GB7FR	DVU40	IO90ST	Worthing
GB7FU	DVU13	IO92TU	Pointon
GB7FW	DVU53	IO92BL	Birmingham
GB7GB	DVU34	IO92AN	Birmingham
GB7GC	DVU35	IO93WN	Grimsby
GB7GD	RV55	IO87VE	Aberdeen
GB7GF	DVU55	IO91RF	Guildford
GB7GG	DVU50	IO85AV	Airdrie
GB7GT	DVU33	IO82QJ	Ludlow
GB7HA	DVU46	JO01HW	Halstead
GB7HB	DVU50	IO64TI	Tandragee
GB7HE	DVU43	JO01QP	Hastings
GB7HF	RB04	IO91VR	Welham Green
GB7HM	DVU51	IO83LC	Caergwrle
GB7HR	DVU36	IO91SL	Heathrow
GB7HS	DVU34	IO93DQ	Cleckheaton
GB7HU	DVU39	IO93RS	South Cave
GB7HX	DVU46	IO93WH	Huddersfield
GB7IC-A	23CM	JO01NI	Herne Bay
GB7IC-B	DVU36	JO01NI	Herne Bay
GB7IC-C	RV53	JO01NI	Herne Bay
GB7IE	RV54	IO70WJ	Plymouth
GB7IK	DVU43	JO01FJ	Rochester
GB7IN	DVU52	IO93GD	Alfreton
GB7IP	RV61	IO92NO	Leicester
GB7IQ	DVU49	IO82WS	Stafford
GB7IS	DVU43	IO81MH	Weston-S-Mare
GB7IT	DVU41	IO81MH	Weston-S-Mare
GB7IV	RV62	IO90HX	Chandlers Ford
GB7JB	DVU37	IO81VC	Mere Wiltshire
GB7JD	DVU33	IO85RL	Jedburgh
GB7JF	DVU51	IO92NE	Northampton
GB7JL	RU77	IO83QL	Wigan
GB7JM	RU71	JO03BI	Louth
GB7KA	DVU57	IO64RW	Kilrea
GB7KB	DVU34	IO91VR	Welham Green
GB7KE	RU78	IO92PJ	Kettering
GB7KH	DVU49	JO01DQ	Kelvedon Hatch
GB7KM	DVU53	IO81XQ	Cirencester
GB7KT	DVU40	IO91GE	Andover
GB7LE	DVU53	IO93FU	Leeds
GB7LF	DVU43	IO84OA	Lancaster
GB7LN	DVU32	IO93RF	Lincoln
GB7LO	DVU41	JO01BJ	Bromley
GB7LP	DVU32	IO83MK	Liverpool
GB7LR	DVU38	IO92IQ	Leicester
GB7LY	DVU53	IO64IX	Derry/Londonderry
GB7MA	RV48	IO83UO	Bury
GB7MB	DVU56	IO84NA	Heysham
GB7MC	DVU43	IO70OJ	St Austell
GB7ME	DVU38	IO92JJ	Rugby
GB7MH	DVU51	IO91WC	East Grinstead
GB7MJ	DVU51	IO91GA	Romsey
GB7MK	DVU48	JO02OB	Ipswich
GB7MR	DVU59	IO83XN	Oldham
GB7MW	DVU39	IO74BS	Carrickfergus
GB7NB	DVU55	JO02PN	Norwich
GB7ND	DVU33	JO02LM	Great Ellingham
GB7NE	DVU36	IO95FE	Ashington
GB7NF	DVU39	JO00AS	Newhaven
GB7NI	RV60	IO74BS	Carrickfergus
GB7NL	DVU37	IO80FR	Exeter
GB7NO	DVU41	IO92MF	Northampton
GB7NS	DVU13	IO91WG	Caterham
GB7NU	DVU43	JO02OW	Sheringham
GB7NY	DVU55	IO64UE	Newry
GB7OK	RV57	JO01BJ	Bromley
GB7OZ	DVU50	IO82KU	Oswestry
GB7PB	RV53	IO94FR	Durham
GB7PD-B	RB05	IO71MQ	Pembroke Dock
GB7PD-C	RV59	IO71MQ	Pembroke Dock
GB7PE	DVU40	IO92XO	Peterborough
GB7PI	DVU61	IO92XA	Royston
GB7PK	DVU42	IO90LT	Portsmouth
GB7PN	DVU34	IO83HH	Prestatyn
GB7PP	DVU35	JO02MG	Ipswich
GB7PT	DVU57	IO92XA	Royston
GB7RB-C	RV54	IO81GK	Cowbridge
GB7RE	DVU57	IO93MH	Retford
GB7RN	RV51	IO90JT	Fareham
GB7RR	DVU48	IO93KA	Nottingham
GB7RV	DVU50	IO83ST	Blackburn
GB7RW	RV48	IO94SO	Whitby
GB7RY	RU76	JO00HW	Rye
GB7SB	RV48	IO90WT	Brighton
GB7SC	DVU38	IO90QT	Bognor Regis
GB7SD	DVU33	IO80SQ	Weymouth
GB7SE	DVU38	JO01DM	Thurrock
GB7SF	RV59	IO93GK	Sheffield
GB7SH	RU77	IO93GK	Sheffield
GB7SI	DVU42	IO83WA	Stoke-On-Trent
GB7SJ	RB07	IO83RF	Northwich
GB7SK	DVU37	IO92NO	Leicester
GB7SN	DVU54	IO93GH	Sheffield
GB7SR	DVU51	IO93GK	Sheffield
GB7SU	DVU36	IO90HV	Southampton
GB7SX	DVU62	IO90QT	Bognor Regis
GB7TC	DVU42	IO91DL	Swindon
GB7TD	DVU13	IO93EP	Wakefield
GB7TE	RV62	JO01OT	Clacton On Sea
GB7TH	DVU49	IO91RI	Broadstairs
GB7TP	DVU55	IO93CT	Shipley
GB7TQ	RU71	IO80FM	Torquay
GB7TT	DVU36	IO75QN	Troon
GB7TV	DVU38	IO94LN	New Marske
GB7TY	DVU49	IO84XX	Hexham
GB7UL	DVU42	IO74BS	Carrickfergus
GB7UZ	DVU40	IO84OB	Lancaster
GB7VO	DVU56	IO82PE	Leominster
GB7WB	DVU39	IO81MH	Weston-S-Mare
GB7WC	DVU39	IO83QI	Warrington
GB7WF	DVU39	IO82UJ	Bewdley
GB7WI	DVU33	IO93RT	Walkington
GB7WL	DVU37	IO91QP	Amersham
GB7WP	DVU61	IO83KJ	Birkenhead
GB7WT	DVU48	IO64HQ	Omagh
GB7WX	RU70	IO83OE	Tarvin
GB7XX	DVU37	IO94FW	Felling
GB7YD-A	23CM	IO93HO	Barnsley
GB7YD-A	RM12	IO93HO	Barnsley
GB7YD-B	DVU41	IO93HO	Barnsley
GB7YD-C	RV54	IO93HO	Barnsley
GB7YR	DVU42	IO93JL	Doncaster
GB7YS	DVU37	IO80QW	Yeovil
GB7YZ	DVU38	IO83JE	Mold
GB7ZI	DVU55	IO82XR	Stafford
GB7ZP	DVU39	JO01GQ	Chelmsford
GB7ZZ	2M	JO01NI	Herne Bay

Repeaters (by output frequency)

Channel	F (out)	F (in)	Callsign	CTCSS	Location	Keeper
10M	29.21	50.52	GB3WX	77.0Hz	Mere Wiltshire	G3ZXX
10M	29.64	29.54	GB3CJ		Northampton	G1IRG
10M	29.69	29.59	GB3GX	103.5Hz	Ludlow	G1MAW
6M	50.52	29.21	GB3WX	77.0Hz	Mere Wiltshire	G3ZXX
R50-01	50.72	51.22	GB3EF	110.9Hz	Stowmarket	G1YFF
R50-02	50.73	51.23	GB3GC	77.0Hz	Gunnislake	M0YDW
R50-02	50.73	51.23	GB3SL	103.5Hz	Kilsyth	GM4COX
R50-02	50.73	51.23	GB3XD	71.9Hz	Louth	G7AJP
R50-03	50.74	51.24	GB3UM		Markfield	M1NAS
R50-04	50.75	51.25	GB3LP	77.0Hz	Liverpool	M1SWB
R50-05	50.76	51.26	GB3HF	71.9Hz	Barnsley	G4LUE
R50-06	50.77	51.27	GB3DB	110.9Hz	Danbury	G6JYB
R50-06	50.77	51.27	GB3FH	77.0Hz	Somerset	G4RKY
R50-07	50.78	51.28	GB3PX	77.0Hz	Royston	G4NBS
R50-07	50.78	51.28	GB3TY	110.9Hz	Carrickfergus	GI6DKQ
R50-08	50.79	51.29	GB3SX	103.5Hz	Stoke On Trent	G4SCY
R50-09	50.8	51.3	GB3WY	82.5Hz	Wakefield	G1XCC
R50-09	50.8	51.3	GB3ZY	77.0Hz	Bristol	G4RKY
R50-10	50.81	51.31	GB3FX	82.5Hz	Farnham	G4EPX
R50-10	50.81	51.31	GB3ZW	103.5Hz	Newtown Powys	GW4IQP
R50-11	50.82	51.32	GB3HM	71.9Hz	Belper	G8IQP
R50-12	50.83	51.33	GB3EC	110.9Hz	Bury St Edmunds	G1YFF
R50-12	50.83	51.33	GB3GT	103.5Hz	Ludlow	G1MAW
R50-13	50.84	51.34	GB3AM	77.0Hz	Amersham	G0RDI
R50-15	50.86	51.36	GB3VI	67.0Hz	Birmingham	G8NDT
2M	145	145	GB7ZZ		Herne Bay	G4TKR
RV48	145.6	145	GB3AS	77.0Hz	Langholm	GM6LJE
RV48	145.6	145	GB3CF	77.0Hz	Markfield	G4AFJ
RV48	145.6	145	GB3EL	77.0Hz	London	G4RZZ
RV48	145.6	145	GB3FF	103.5Hz	Kelty	GM7LUN
RV48	145.6	145	GB3LY	110.9Hz	Limavady	GI3USS
RV48	145.6	145	GB3NZ	94.8Hz	Thorpe St Andrew	M0ZAH
RV48	145.6	145	GB3SS	67.0Hz	Elgin Moray	GM4ILS
RV48	145.6	145	GB3WR	94.8Hz	Cheddar	G4RKY
RV48	145.6	145	GB7MA		Bury	G7LWT
RV48	145.6	145	GB7RW	88.5Hz	Whitby	G4EQS
RV48	145.6	145	GB7SB		Brighton	G4PAP
RV49	145.6125	145.0125	GB3EI	88.5Hz	Clachan North Uist	G8SAU
RV49	145.6125	145.0125	GB3EW	77.0Hz	Exeter	G8XQQ
RV49	145.6125	145.0125	GB3NA	71.9Hz	Barnsley	G4LUE
RV49	145.6125	145.0125	GB3OA	82.5Hz	Southport	G4EID
RV49	145.6125	145.0125	GB3TE	103.5Hz	Clacton On Sea	G0MBA
RV49	145.6125	145.0125	GB3TO	77.0Hz	Northampton	G6NYH
RV49	145.6125	145.0125	GB3VM	103.5Hz	Ludlow	G4AIJ
RV50	145.625	145.025	GB3GD	110.9Hz	Snaefell Iom	GD4HOZ
RV50	145.625	145.025	GB3HG	88.5Hz	Thirsk	G8IMZ
RV50	145.625	145.025	GB3KS	103.5Hz	Dover	M1CMN
RV50	145.625	145.025	GB3MH	88.5Hz	East Grinstead	G3NZP
RV50	145.625	145.025	GB3NB	94.8Hz	Norwich	G8VLL
RV50	145.625	145.025	GB3NF	77.0Hz	Nottingham	G4NRZ
RV50	145.625	145.025	GB3NG	67.0Hz	Fraserburgh	GM4ZUK
RV50	145.625	145.025	GB3NW	67.0Hz	Worcester	G4IDF
RV50	145.625	145.025	GB3PA	103.5Hz	Paisley	GM7OAW
RV50	145.625	145.025	GB3SI		Helston	M1ERD
RV51	145.6375	145.0375	GB3DN	77.0Hz	Stibb Cross	G1BHM
RV51	145.6375	145.0375	GB3GJ	71.9Hz	St Helier	GJ8PVL
RV51	145.6375	145.0375	GB3IN	71.9Hz/DMR1	Alfreton	G4TSN
RV51	145.6375	145.0375	GB3RA	103.5Hz	Llandrindod Wells	GW7UNV
RV51	145.6375	145.0375	GB3VO	110.9Hz	Chirk	GW0WZZ
RV51	145.6375	145.0375	GB7CT	DMR/3	Tring	G0RDI
RV51	145.6375	145.0375	GB7DE	DMR/1	Edinburgh	GM7RYR
RV51	145.6375	145.0375	GB7RN		Fareham	G3ZDF
RV52	145.65	145.05	GB3AY	103.5Hz	Dalry	MM0YET
RV52	145.65	145.05	GB3HS	88.5Hz	Walkington	G3GJA
RV52	145.65	145.05	GB3MN	82.5Hz	Disley	G8LZO
RV52	145.65	145.05	GB3OC	77.0Hz	Kirkwall	GM0HQG
RV52	145.65	145.05	GB3PO	110.9Hz	Ipswich	G7CIY
RV52	145.65	145.05	GB3SB	118.8Hz	Selkirk	GM0FTJ
RV52	145.65	145.05	GB3TR	94.8Hz	Torquay	G8XST
RV52	145.65	145.05	GB3WH	118.8Hz	Swindon	G4LDL
RV53	145.6625	145.0625	GB3AA	94.8Hz	Bristol	G4CJZ
RV53	145.6625	145.0625	GB3CO	77.0Hz	Corby	G1DIW
RV53	145.6625	145.0625	GB3DW	110.9Hz	Harlech	MW0VTK
RV53	145.6625	145.0625	GB3FE	103.5Hz	Stirling	GM0MZB
RV53	145.6625	145.0625	GB3KI	103.5Hz	Herne Bay	G4TKR
RV53	145.6625	145.0625	GB3PK	110.9Hz	Ballycastle Ni	MI0CRR
RV53	145.6625	145.0625	GB3SH	71.9Hz	Romsey	G4MYS
RV53	145.6625	145.0625	GB7IC-C		Herne Bay	G4TKR
RV53	145.6625	145.0625	GB7PB		Durham	G4EBN
RV54	145.675	145.075	GB3BX	146.2Hz	Much Wenlock	M1GIZ
RV54	145.675	145.075	GB3ES	103.5Hz	Hastings	G6ZZX
RV54	145.675	145.075	GB3LD	110.9Hz	Lancaster	G3VVT
RV54	145.675	145.075	GB3LU	77.0Hz	Lerwick	GM4SLV
RV54	145.675	145.075	GB3PE	94.8Hz	Peterborough	M0ZPU
RV54	145.675	145.075	GB3PR	94.8Hz	Perth	GM8KPH
RV54	145.675	145.075	GB3RD	118.8Hz	Reading	G8DOR
RV54	145.675	145.075	GB7IE	77.0Hz	Plymouth	G7DIR
RV54	145.675	145.075	GB7RB-C		Cowbridge	GW6CUR
RV54	145.675	145.075	GB7YD-C		Barnsley	G4LUE
RV55	145.6875	145.0875	GB3CD	118.8Hz	Crook	G0OCB
RV55	145.6875	145.0875	GB3DC	71.9Hz	Derby	G7NPW
RV55	145.6875	145.0875	GB3EA	110.9Hz	Wickhambrook	G1YFF
RV55	145.6875	145.0875	GB3KE	103.5Hz	Glasgow	GM7SVK
RV55	145.6875	145.0875	GB3WE	94.8Hz/DMR1	Weston-S-Mare	G4SZM
RV55	145.6875	145.0875	GB3XP	82.5Hz	Morden	M0SGL
RV55	145.6875	145.0875	GB7DK	103.5Hz	Stranraer	GM0HPK
RV55	145.6875	145.0875	GB7GD		Aberdeen	MM0CUG
RV56	145.7	145.1	GB3AR	110.9Hz	Caernarfon	GW4KAZ
RV56	145.7	145.1	GB3BB	94.8Hz	Brecon	MW0UAA
RV56	145.7	145.1	GB3BT	118.8Hz	Berwick On Tweed	GM1JFF
RV56	145.7	145.1	GB3EV	77.0Hz	Dufton	G7ITT
RV56	145.7	145.1	GB3HH	71.9Hz	Buxton	G7EKY
RV56	145.7	145.1	GB3HI	103.5Hz	Isle Of Mull	MM0JRM
RV56	145.7	145.1	GB3KN	94.8Hz	Maidstone	G3VFC
RV56	145.7	145.1	GB3VA	118.8Hz	Brill	G8BQH
RV56	145.7	145.1	GB3WD	77.0Hz	Plymouth	G7LUL
RV56	145.7	145.1	GB7CM	71.9Hz	Blandford	M0MRP
RV57	145.7125	145.1125	GB3BM	67.0Hz	Birmingham	G8AMD
RV57	145.7125	145.1125	GB3FG	94.8Hz	Carmarthen	GW8KCY
RV57	145.7125	145.1125	GB3KY	94.8Hz	Kings Lynn	G1SCQ
RV57	145.7125	145.1125	GB3LA	103.5Hz	Sanquhar	GM3SAN
RV57	145.7125	145.1125	GB3MI	82.5Hz	Manchester	G0TOG
RV57	145.7125	145.1125	GB3SW	77.0Hz	Sidmouth	G6XUV
RV57	145.7125	145.1125	GB7OK		Bromley	G1HIG
RV58	145.725	145.125	GB3AG	94.8Hz	Forfar	GM1CMF
RV58	145.725	145.125	GB3BI	67.0Hz	Inverness	GM1VAD
RV58	145.725	145.125	GB3CG	118.8Hz	Gloucester	G3LVP
RV58	145.725	145.125	GB3DA	110.9Hz	Danbury	G6JYB
RV58	145.725	145.125	GB3LM	71.9Hz	Lincoln	G7AVU
RV58	145.725	145.125	GB3NC	77.0Hz	Roche	G4WVD
RV58	145.725	145.125	GB3NI	110.9Hz	Belfast N.I.	GI3USS
RV58	145.725	145.125	GB3SN	71.9Hz	Alton Hants	G4EPX
RV58	145.725	145.125	GB3TP	82.5Hz	Shipley	G8ZMG
RV58	145.725	145.125	GB3TW	118.8Hz	Gateshead	G7UUR
RV58	145.725	145.125	GB3VT	103.5Hz	Stoke On Trent	G8NSS
RV59	145.7375	145.1375	GB3AL	77.0Hz	Amersham	G0RDI
RV59	145.7375	145.1375	GB3CP	110.9Hz	Fermanagh	GI8RLE
RV59	145.7375	145.1375	GB3DR	71.9Hz	Dorchester	G6WHI
RV59	145.7375	145.1375	GB3TA	67.0Hz	Tamworth	M0TSD
RV59	145.7375	145.1375	GB3ZA	118.8Hz	Hereford	G0JWJ
RV59	145.7375	145.1375	GB7PD-C		Pembroke Dock	MW0XDT
RV59	145.7375	145.1375	GB7SF		Sheffield	M1ERS
RV60	145.75	145.15	GB3CS	103.5Hz	Motherwell	GM8HBY
RV60	145.75	145.15	GB3FK	103.5Hz	Folkestone	M1CMN
RV60	145.75	145.15	GB3HA	118.8Hz	Corbridge	G7UUR
RV60	145.75	145.15	GB3IS	88.5Hz	Broadford	GM8RBR
RV60	145.75	145.15	GB3MP	110.9Hz	Denbigh	M0OBW
RV60	145.75	145.15	GB3PI	77.0Hz	Royston	G4NBS
RV60	145.75	145.15	GB3WS	88.5Hz	Crawley Sussex	G4EFO
RV60	145.75	145.15	GB3YC	88.5Hz	Driffield	M0KXQ
RV60	145.75	145.15	GB7NI	DMR/1	Carrickfergus	GI6DKQ
RV61	145.7625	145.1625	GB3DU	118.8Hz	Duns	GM7LUN
RV61	145.7625	145.1625	GB3IK	103.5Hz	Rochester	G6CKK
RV61	145.7625	145.1625	GB3IP	DMR/1	Stafford	G7PFT
RV61	145.7625	145.1625	GB3IR	88.5Hz	Richmond Yorks	G4FZN
RV61	145.7625	145.1625	GB3NE	118.8Hz	Newbury	G6IBI
RV61	145.7625	145.1625	GB3PL	77.0Hz	E.Cornwall	M0YDW
RV61	145.7625	145.1625	GB7IP	77.0Hz/DMR1	Leicester	M1FJB
RV62	145.775	145.175	GB3DG	103.5Hz	Newton Stewart	MM1BHO
RV62	145.775	145.175	GB3FR	71.9Hz	Spilsby Lincs.	M1FJB
RV62	145.775	145.175	GB3GN	67.0Hz	Banchory	GM4ZUK
RV62	145.775	145.175	GB3IG	88.5Hz	Stornoway	GM0LZE
RV62	145.775	145.175	GB3NL		North London	G4DFB
RV62	145.775	145.175	GB3PW	103.5Hz	Newtown Powys	GW4IQP
RV62	145.775	145.175	GB3RF	82.5Hz/DMR2	Accrington	G0BMH
RV62	145.775	145.175	GB3WK		Leamington Spa	G6FEO
RV62	145.775	145.175	GB3WT	110.9Hz	Omagh	GI3NVW
RV62	145.775	145.175	GB3WW	94.8Hz	Port Talbot	GW4FOI
RV62	145.775	145.175	GB7DA		Airdrie	GM4AUP
RV62	145.775	145.175	GB7ER	77.0Hz	Exeter	M0ZZT
RV62	145.775	145.175	GB7IV		Chandlers Ford	G4MYS
RV62	145.775	145.175	GB7TE		Clacton On Sea	G0MBA
RV63	145.7875	145.1875	GB3BF	77.0Hz	Bedford	G8MGP
RV63	145.7875	145.1875	GB3EB	88.5Hz	Uckfield	G8PUO
RV63	145.7875	145.1875	GB3GO		Llandudno	MW0TMH
RV63	145.7875	145.1875	GB3JB	103.5Hz	Mere Wiltshire	G3ZXX
RV63	145.7875	145.1875	GB3JL	77.0Hz	Liskeard	G4RKY
RV63	145.7875	145.1875	GB3KD	118.8Hz	Kidderminster	G8PZT
RV63	145.7875	145.1875	GB3LB	118.8Hz	Lauder	GM7LUN
RV63	145.7875	145.1875	GB3YW	82.5Hz	Wakefield	G1XCC
RV63	145.7875	145.1875	GB7FK-C		Folkestone	M1CMN
RU66	430.825	438.425	GB3CI	77.0Hz	Corby	G7HPE
RU66	430.825	438.425	GB3IM	110.9Hz	Douglas Iom	GD4HOZ
RU66	430.825	438.425	GB3IM	71.9Hz	Ramsey Iom	GD4HOZ
RU66	430.825	438.425	GB3KW	103.5Hz	Glasgow	GM7SVK
RU66	430.825	438.425	GB3NY	88.5Hz	Bridlington	M0DPH
RU66	430.825	438.425	GB3UO	110.9Hz	Chirk	GW0WZZ
RU66	430.825	438.425	GB3WU	118.8Hz	Worcester	G8TIC

Repeaters (by output frequency)

Channel	F (out)	F (in)	Callsign	CTCSS	Location	Keeper
RU66	430.825	438.425	GB3ZB	77.0Hz	Bristol	G4RKY
RU66	430.825	438.425	GB3ZX	88.5Hz	Eastbourne	G6ZZX
RU68	430.85	438.45	GB3BS	118.8Hz	Bristol	G4SDR
RU68	430.85	438.45	GB3BZ	110.9Hz	Braintree	G0DEC
RU68	430.85	438.45	GB3FC	82.5Hz	Blackpool	G6AOS
RU68	430.85	438.45	GB3GR	71.9Hz	Grantham	G8SAU
RU68	430.85	438.45	GB3IE	77.0Hz	Plymouth	G7DQC
RU68	430.85	438.45	GB3NX	88.5Hz	Crawley Sussex	G4EFO
RU69	430.8625	438.4625	GB3BE	118.8Hz	Duns	GM7LUN
RU69	430.8625	438.4625	GB3LR	88.5Hz	Newhaven	G0TJH
RU69	430.8625	438.4625	GB3UK	103.5Hz	Cheltenham	G0LGS
RU69	430.8625	438.4625	GB3UK	103.5Hz	Cheltenham	G0LGS
RU69	430.8625	438.4625	GB7DO	82.5Hz/DMR5	Doncaster	G1ILF
RU70	430.875	438.475	GB3IM	110.9Hz	Peel	GD4HOZ
RU70	430.875	438.475	GB3NH	77.0Hz	Northampton	G4YKE
RU70	430.875	438.475	GB3WA	103.5Hz	Dalry	GM7GDE
RU70	430.875	438.475	GB3WG	94.8Hz	Port Talbot	GW4FOI
RU70	430.875	438.475	GB7WX	103.5Hz	Tarvin	G7NEH
RU71	430.8875	438.4875	GB3BP		Belper	G0MGX
RU71	430.8875	438.4875	GB3EK	103.5Hz	Margate	M0LMK
RU71	430.8875	438.4875	GB3HO	88.5Hz	Horsham	G4EFO
RU71	430.8875	438.4875	GB3NP	77.0Hz	Towcester	G4YKE
RU71	430.8875	438.4875	GB3NU	94.8Hz	Sheringham	G8SAU
RU71	430.8875	438.4875	GB3XL	82.5Hz/DMR1	Shipley	M0IRK
RU71	430.8875	438.4875	GB7FI	77.0Hz/DMR3	Axbridge	G4RKY
RU71	430.8875	438.4875	GB7JM	DMR/3	Louth	M0AQC
RU71	430.8875	438.4875	GB7TQ	94.8Hz	Torquay	G8XST
RU72	430.9	438.5	GB3GL	103.5Hz	Glasgow	GM3SAN
RU72	430.9	438.5	GB3HY		Haywards Heath	G6DGK
RU72	430.9	438.5	GB3JC	94.8Hz	Norwich	M0ZAH
RU72	430.9	438.5	GB3JU	88.5Hz	St Helier	GJ8PVL
RU72	430.9	438.5	GB3KC	67.0Hz	Stourbridge	G0EWH
RU72	430.9	438.5	GB3LW	82.5Hz	London	G4RFC
RU72	430.9	438.5	GB3MT	110.9Hz	Magherafelt	MI0GRN
RU72	430.9	438.5	GB3PZ	82.5Hz	Dukinfield	G4ZPZ
RU73	430.9125	438.5125	GB3XX	77.0Hz	Daventry	G8KHF
RU74	430.925	438.525	GB3DM	103.5Hz	Dumbarton	GM7GDE
RU74	430.925	438.525	GB3FI	77.0Hz/DMR3	Cheddar	G4RKY
RU74	430.925	438.525	GB3VN	103.5Hz	Ludlow	G4OYX
RU74	430.925	438.525	GB3XN	71.9Hz	Worksop	G3XXN
RU74	430.925	438.525	GB7BR	DMR/3	Ramsey Iom	GD4HOZ
RU74	430.925	438.525	GB7CA	DMR/2	Douglas	GD4HOZ
RU75	430.9375	438.5375	GB3SF	77.0Hz	South Molton	G6SQX
RU75	430.9375	438.5375	GB3WP	82.5Hz	Hyde	G6YRK
RU76	430.95	438.55	GB3EX	77.0Hz	Silverton	G7NBU
RU76	430.95	438.55	GB3FJ	71.9Hz	Asgarby	G3ZPU
RU76	430.95	438.55	GB3OH	94.8Hz	Linlithgow	GM0MZB
RU76	430.95	438.55	GB3OM	110.9Hz/DMR1	Omagh	GI4SXV
RU76	430.95	438.55	GB3OY	77.0Hz	Buckhurst Hill	G7UZN
RU76	430.95	438.55	GB3WL	67.0Hz	Birmingham	G3YXM
RU76	430.95	438.55	GB7BD	DMR/3	Bristol	G4RKY
RU76	430.95	438.55	GB7FG	94.8Hz	Carmarthen	GW8KCY
RU76	430.95	438.55	GB7RY	103.5Hz	Rye	M0HOW
RU77	430.9625	438.5625	GB3CC	88.5Hz	Chichester	G3UEQ
RU77	430.9625	438.5625	GB7JL	82.5Hz/DMR1	Wigan	G1EFU
RU77	430.9625	438.5625	GB7SH	71.9Hz	Sheffield	M1ERS
RU78	430.975	438.575	GB3AC	94.8Hz	Lydney Glos	G4CJZ
RU78	430.975	438.575	GB3DQ	77.0Hz	Polperro	G1YDQ
RU78	430.975	438.575	GB3EZ	110.9Hz	Wickhambrook	G1YFF
RU78	430.975	438.575	GB3KK	110.9Hz	Ballycastle	MI0CRQ
RU78	430.975	438.575	GB3KV	103.5Hz	Kilsyth	GM3SAN
RU78	430.975	438.575	GB3WO	88.5Hz	Lancing	G1VUP
RU78	430.975	438.575	GB3ZI	77.0Hz	Stafford	G4YFF
RU78	430.975	438.575	GB7FC	82.5Hz/DMR1	Blackpool	M0AUT
RU78	430.975	438.575	GB7KE		Kettering	G3XFA
RB0	433	434.6	GB3BN	118.8Hz	Bracknell	G8DOR
RB0	433	434.6	GB3CK	103.5Hz	Charing Kent	M0ZAA
RB0	433	434.6	GB3DT	71.9Hz	Blandford	G0ZEP
RB0	433	434.6	GB3LL	110.9Hz	Llandudno	GW6SIX
RB0	433	434.6	GB3MK	77.0Hz	Milton Keynes	G4CAK
RB0	433	434.6	GB3NR	94.8Hz	Norwich	G8VLL
RB0	433	434.6	GB3NT	118.8Hz	Newcastle Upon Tyne	G7UUR
RB0	433	434.6	GB3PU	94.8Hz	Perth	GM8KPH
RB0	433	434.6	GB3RT	94.8Hz	Cwmbran	MW0YAC
RB0	433	434.6	GB3US	103.5Hz	Sheffield	G4CUI
RB0	433	434.6	GB3WN	67.0Hz	Wolverhampton	G4OKE
RB01	433.025	434.625	GB3BV	118.8Hz	Hemel Hempstead	G3YXZ
RB01	433.025	434.625	GB3DV	71.9Hz	Maltby	G0EPX
RB01	433.025	434.625	GB3HJ	118.8Hz	Harrogate	G4MEM
RB01	433.025	434.625	GB3JS	94.8Hz	Great Yarmouth	M0JGX
RB01	433.025	434.625	GB3MA	82.5Hz	Bury	G7LWT
RB02	433.05	434.65	GB3AV	118.8Hz	Aylesbury	G8BQH
RB02	433.05	434.65	GB3CH	77.0Hz	Liskeard	G4RKY
RB02	433.05	434.65	GB3HK	118.8Hz	Selkirk	GM0FTJ
RB02	433.05	434.65	GB3LS	71.9Hz	Lincoln	G7AVU
RB02	433.05	434.65	GB3LV	82.5Hz	North London	G4DFB
RB02	433.05	434.65	GB3MF	103.5Hz	Macclesfield	G1JVF
RB02	433.05	434.65	GB3NN	94.8Hz	Wells Norfolk	G0FVF
RB02	433.05	434.65	GB3UL	110.9Hz	Belfast N.I.	GI3USS
RB02	433.05	434.65	GB3YS	88.5Hz	Yeovil	G3ZXX
RB03	433.075	434.675	GB3ER	110.9Hz	Danbury	G6JYB
RB03	433.075	434.675	GB3KR	67.0Hz	Kidderminster	G8NTU
RB03	433.075	434.675	GB3KU	82.5Hz	Ashton-U-Lyne	M0NCZ
RB03	433.075	434.675	GB3TD	118.8Hz	Swindon	G4XUT
RB03	433.075	434.675	GB3VS	94.8Hz	Taunton	G4UVZ
RB04	433.1	434.7	GB3CW	103.5Hz	Newtown Powys	GW4IQP
RB04	433.1	434.7	GB3IH	110.9Hz	Ipswich	G7CIY
RB04	433.1	434.7	GB3KL	94.8Hz	Kings Lynn	G0IJU
RB04	433.1	434.7	GB3LE	77.0Hz	Markfield	M1NAS
RB04	433.1	434.7	GB3NK	103.5Hz	Erith	G4EGU
RB04	433.1	434.7	GB3SP	94.8Hz	Pembroke	GW4VRO
RB04	433.1	434.7	GB3UB	118.8Hz	Bath Avon	G4KVI
RB04	433.1	434.7	GB3VE	77.0Hz	Dufton	G7ITT
RB04	433.1	434.7	GB7HF	82.5Hz	Welham Green	G1YJH
RB05	433.125	434.725	GB3GH	118.8Hz	Gloucester	G3LVP
RB05	433.125	434.725	GB3IC	67.0Hz	Wolverhampton	M0VRR
RB05	433.125	434.725	GB3IM	110.9Hz	Douglas Iom	GD4HOZ
RB05	433.125	434.725	GB3OV	94.8Hz	St Neots	M1JUL
RB05	433.125	434.725	GB3WJ	88.5Hz	Scunthorpe	G3TMD
RB05	433.125	434.725	GB7PD-B		Pembroke Dock	MW0XDT
RB06	433.15	434.75	GB3BR	88.5Hz	Brighton	G4PAP
RB06	433.15	434.75	GB3CR	110.9Hz	Caergwrle	M0OBW
RB06	433.15	434.75	GB3DI	118.8Hz	Didcot	G8CUL
RB06	433.15	434.75	GB3HC	118.8Hz	Hereford	G0JWJ
RB06	433.15	434.75	GB3ME	67.0Hz	Rugby	G7BQM
RB06	433.15	434.75	GB3SK	103.5Hz	Canterbury	G6DIK
RB06	433.15	434.75	GB3SY	71.9Hz	Barnsley	G4LUE
RB07	433.175	434.775	GB3AB	82.5Hz	Sheffield	M0GAV
RB07	433.175	434.775	GB3AU	82.5Hz	Amersham	G0RDI
RB07	433.175	434.775	GB3BL	77.0Hz	Bedford	G8MGP
RB07	433.175	434.775	GB3DE	110.9Hz	Ipswich	G1NRL
RB07	433.175	434.775	GB3MS	77.0Hz	Worcester	G7WIG
RB07	433.175	434.775	GB3NM	71.9Hz	Nottingham	G4IRX
RB07	433.175	434.775	GB3TS	118.8Hz	Sunderland	G7MFN
RB07	433.175	434.775	GB7SJ	103.5Hz	Northwich	M0WTX
RB08	433.2	434.8	GB3AN	110.9Hz	Amlwch	GW6DOK
RB08	433.2	434.8	GB3CM	94.8Hz	Carmarthen	GW8KCY
RB08	433.2	434.8	GB3EH	67.0Hz	Edge Hill	G4OHB
RB08	433.2	434.8	GB3PY	77.0Hz	Cambridge	G4NBS
RB08	433.2	434.8	GB3RB	71.9Hz	Bolsover	G1SLE
RB08	433.2	434.8	GB7TF	103.5Hz	Telford	G3UKV
RB09	433.225	434.825	GB3BG	94.8Hz	Blaenavon	GW7LOP
RB09	433.225	434.825	GB3CL	103.5Hz	Clacton	G0MBA
RB09	433.225	434.825	GB3CV	67.0Hz	Coventry	G7TRJ
RB09	433.225	434.825	GB3HD	82.5Hz	Huddersfield	G0ISX
RB09	433.225	434.825	GB3IW	71.9Hz	Ryde	G4IKI
RB09	433.225	434.825	GB3LC	71.9Hz	Louth	G7AJP
RB09	433.225	434.825	GB3ST	103.5Hz	Stoke On Trent	G8NSS
RB09	433.225	434.825	GB3TU	77.0Hz	Tring Herts	G0RDI
RB10	433.25	434.85	GB3AW	71.9Hz	Newbury	G8DOR
RB10	433.25	434.85	GB3DD	94.8Hz	Dundee	GM4UGF
RB10	433.25	434.85	GB3DY	71.9Hz	Wirksworth	G3ZYC
RB10	433.25	434.85	GB3LI	82.5Hz	Liverpool	G3WIC
RB10	433.25	434.85	GB3LT	77.0Hz	Luton	G8XTW
RB10	433.25	434.85	GB3MW		Leamington Spa	G6FEO
RB11	433.275	434.875	GB3AH	94.8Hz	East Dereham	G0LGJ
RB11	433.275	434.875	GB3GY	88.5Hz	Cleethorpes	M0KWK
RB11	433.275	434.875	GB3HT	77.0Hz	Hinckley	G4ALB
RB11	433.275	434.875	GB3RE	103.5Hz	Maidstone	G6RVS
RB11	433.275	434.875	GB3RH	94.8Hz	Axminster	G6WWY
RB11	433.275	434.875	GB3RU	118.8Hz	Reading	G8DOR
RB12	433.3	434.9	GB3DX	110.9Hz	Derry/Londonderry	GI4YWT
RB12	433.3	434.9	GB3EE	71.9Hz	Chesterfield	G1SLE
RB12	433.3	434.9	GB3GB	67.0Hz	Birmingham	G8NDT
RB12	433.3	434.9	GB3GF	88.5Hz	Guildford	G4EML
RB12	433.3	434.9	GB3TJ	118.8Hz	Corbridge	G7UUR
RB12	433.3	434.9	GB3WB	94.8Hz	Weston-S-Mare	G4TBD
RB13	433.325	434.925	GB3CA	77.0Hz	Carlisle	G1XSZ
RB13	433.325	434.925	GB3DS	71.9Hz	Worksop	G3XXN
RB13	433.325	434.925	GB3GU	71.9Hz	Guernsey	GU6EFB
RB13	433.325	434.925	GB3HW	110.9Hz	Gidea Park	G4GBW
RB13	433.325	434.925	GB3SM	103.5Hz	Stoke On Trent	G4SCY
RB13	433.325	434.925	GB3VH	82.5Hz	Welwyn Garden City	G4THF
RB14	433.35	434.95	GB3CB	67.0Hz	Birmingham	G8VIQ
RB14	433.35	434.95	GB3CE	110.9Hz	Colchester	G0MBA
RB14	433.35	434.95	GB3ED	94.8Hz	Edinburgh	GM4GZW
RB14	433.35	434.95	GB3HE	103.5Hz	Hastings	G8PUO
RB14	433.35	434.95	GB3HR	82.5Hz	Harrow	G3YXZ
RB14	433.35	434.95	GB3LF	110.9Hz	Kendal	G3VVT
RB14	433.35	434.95	GB3MR	82.5Hz	Stockport	G8LZO
RB14	433.35	434.95	GB3ND	77.0Hz	Bideford	G4SOF
RB14	433.35	434.95	GB3SD	71.9Hz	Weymouth	G0EVW
RB14	433.35	434.95	GB3WF	82.5Hz	Otley	M0SNW
RB14	433.35	434.95	GB3YL	94.8Hz	Lowestoft	G4RKP
RB15	433.375	434.975	GB3FN	82.5Hz	Farnham	G4EPX
RB15	433.375	434.975	GB3HB	77.0Hz	Roche	G4WVD
RB15	433.375	434.975	GB3LH	103.5Hz	Shrewsbury	G8DIR
RB15	433.375	434.975	GB3MB	94.8Hz	Merthyr Tydfil	GW6CUR
RB15	433.375	434.975	GB3PP	82.5Hz	Preston	M0NED

Repeaters (by output frequency)

Channel	F (out)	F (in)	Callsign	CTCSS	Location	Keeper
RB15	433.375	434.975	GB3TH	67.0Hz	Tamworth	G8YUQ
RB15	433.375	434.975	GB3WI	94.8Hz	Wisbech	M0DUQ
DVU13	439.1625	430.1625	GB7BS	DMR/3	Bristol	G4SDR
DVU13	439.1625	430.1625	GB7FU	DMR/1	Pointon	G8SJP
DVU13	439.1625	430.1625	GB7NS	DMR/3	Caterham	G0OLX
DVU13	439.1625	430.1625	GB7TD	DMR/1	Wakefield	G1XCC
DVU32	439.4	430.4	GB7AL	DMR/2	Ipswich	M1NIZ
DVU32	439.4	430.4	GB7BM		Birmingham	G8VIQ
DVU32	439.4	430.4	GB7CW	DMR/3	Bridgend	GW0UZK
DVU32	439.4	430.4	GB7EP	DMR/1	Epsom	G0OXZ
DVU32	439.4	430.4	GB7LN	DMR/1	Lincoln	G0RZR
DVU32	439.4	430.4	GB7LP	DMR/1	Liverpool	M1SWB
DVU33	439.4125	430.4125	GB7DN		Dunginven	GI0AZB
DVU33	439.4125	430.4125	GB7GT	DMR/13	Ludlow	G1MAW
DVU33	439.4125	430.4125	GB7JD	DMR/1	Jedburgh	GM4UPX
DVU33	439.4125	430.4125	GB7ND	DMR/1	Great Ellingham	G0LGJ
DVU33	439.4125	430.4125	GB7SD	DMR/1	Weymouth	G0EVW
DVU33	439.4125	430.4125	GB7WI	DMR/1	Walkington	G0UZJ
DVU34	439.425	430.425	GB7AS	DMR/3	Ashford	M1CMN
DVU34	439.425	430.425	GB7DR	DMR/5	Poole	G7ICH
DVU34	439.425	430.425	GB7DS	94.8Hz/DMR1	Norwich	M0ZAH
DVU34	439.425	430.425	GB7GB	DMR/8	Birmingham	G8NDT
DVU34	439.425	430.425	GB7HS	DMR/2	Cleckheaton	G1XCC
DVU34	439.425	430.425	GB7KB		Welham Green	G1YJH
DVU34	439.425	430.425	GB7PN	DMR/1	Prestatyn	G4NOY
DVU35	439.4375	430.4375	GB7AV	DMR/3	Aylesbury	G0RAS
DVU35	439.4375	430.4375	GB7BX	DMR/5	Much Wenlock	M1GIZ
DVU35	439.4375	430.4375	GB7ES		Eastbourne	M0LRE
DVU35	439.4375	430.4375	GB7FO	DMR/1	Blackpool	G0WDA
DVU35	439.4375	430.4375	GB7GC	DMR/2	Grimsby	G7EOG
DVU35	439.4375	430.4375	GB7PP		Ipswich	G0FEA
DVU36	439.45	430.45	GB7BP		Milton Keynes	M1ACB
DVU36	439.45	430.45	GB7HR	DMR/3	Heathrow	G0OXZ
DVU36	439.45	430.45	GB7IC-B		Herne Bay	G4TKR
DVU36	439.45	430.45	GB7NE		Ashington	G0UDZ
DVU36	439.45	430.45	GB7SU	DMR/8	Southampton	G6IGA
DVU36	439.45	430.45	GB7TT		Troon	GM4XRY
DVU37	439.4625	430.4625	GB7JB	DMR/1	Mere Wiltshire	G3ZXX
DVU37	439.4625	430.4625	GB7NL		Exeter	M0NLO
DVU37	439.4625	430.4625	GB7SK	DMR/1	Leicester	M1FJB
DVU37	439.4625	430.4625	GB7WL	DMR/3	Amersham	G0RDI
DVU37	439.4625	430.4625	GB7XX	DMR/10	Felling	G4MSF
DVU38	439.475	430.475	GB7CH		Louth	G8YAP
DVU38	439.475	430.475	GB7LR	DMR/1	Leicester	M1FJB
DVU38	439.475	430.475	GB7SC	DMR/3	Bognor Regis	G0AFN
DVU38	439.475	430.475	GB7SE	DMR/3	Thurrock	M0PFX
DVU38	439.475	430.475	GB7TV		New Marske	M0RIG
DVU38	439.475	430.475	GB7YZ	110.9Hz	Mold	M0WTX
DVU39	439.4875	430.4875	GB7DC	DMR/1	Derby	G7NPW
DVU39	439.4875	430.4875	GB7HU		South Cave	G0VRM
DVU39	439.4875	430.4875	GB7MW	DMR/15	Carrickfergus	GI6DKQ
DVU39	439.4875	430.4875	GB7NF	DMR/1	Newhaven	G0TJH
DVU39	439.4875	430.4875	GB7WB	DMR/2	Weston-S-Mare	G4SZM
DVU39	439.4875	430.4875	GB7WC		Warrington	G4VSS
DVU39	439.4875	430.4875	GB7WF		Bewdley	G8OXG
DVU39	439.4875	430.4875	GB7ZP		Chelmsford	G6JYB
DVU40	439.5	430.5	GB7DP-B		Dundee	GM0ROU
DVU40	439.5	430.5	GB7FR	DMR/4	Worthing	G7RZU
DVU40	439.5	430.5	GB7KT	DMR/1	Andover	G3ZXX
DVU40	439.5	430.5	GB7PE	DMR/3	Peterborough	M0ZPU
DVU40	439.5	430.5	GB7UZ	DMR/1	Lancaster	G4TUZ
DVU41	439.5125	430.5125	GB7AH	DMR/1	Ahoghill	MI0CUN
DVU41	439.5125	430.5125	GB7CF		Folkestone	M1CMN
DVU41	439.5125	430.5125	GB7IT	DMR/1	Weston-S-Mare	G4SZM
DVU41	439.5125	430.5125	GB7LO	DMR/3	Bromley	G1HIG
DVU41	439.5125	430.5125	GB7NO	DMR/3	Northampton	M0NCW
DVU41	439.5125	430.5125	GB7YD-B		Barnsley	G4LUE
DVU42	439.525	430.525	GB7AK	DMR/3	Barking	G8YPK
DVU42	439.525	430.525	GB7PK	DMR/1	Portsmouth	G7RPG
DVU42	439.525	430.525	GB7SI		Stoke-On-Trent	G8NSS
DVU42	439.525	430.525	GB7TC	DMR/2	Swindon	G8VRI
DVU42	439.525	430.525	GB7UL	DMR/1	Carrickfergus	GI6DKQ
DVU42	439.525	430.525	GB7YR	DMR/2	Doncaster	M1DAH
DVU43	439.5375	430.5375	GB7DG		Bromsgrove	M1JSS
DVU43	439.5375	430.5375	GB7EN		Wellingborough	G7HIF
DVU43	439.5375	430.5375	GB7HE		Hastings	G8PUO
DVU43	439.5375	430.5375	GB7IK	DMR/3	Rochester	G6CKK
DVU43	439.5375	430.5375	GB7IS		Weston-S-Mare	G4SZM
DVU43	439.5375	430.5375	GB7LF		Lancaster	G3VVT
DVU43	439.5375	430.5375	GB7MC		St Austell	M1DNS
DVU43	439.5375	430.5375	GB7NU		Sheringham	G8SAU
DVU46	439.575	430.575	GB7EB	DMR/2	Beccles	M0JGX
DVU46	439.575	430.575	GB7HA	DMR/3	Halstead	M0NAS
DVU46	439.575	430.575	GB7HX	DMR/1	Huddersfield	G0ISX
DVU48	439.6	430.6	GB7MK	DMR/13	Ipswich	M1NIZ
DVU48	439.6	430.6	GB7RR	DMR/1	Nottingham	G0LCG
DVU48	439.6	430.6	GB7WT		Omagh	GI3NVW
DVU49	439.6125	430.6125	GB7DV		St. Helens	G1DVA
DVU49	439.6125	430.6125	GB7FH		Fareham	G6ORL
DVU49	439.6125	430.6125	GB7IQ	DMR/1	Stafford	G7PFT
DVU49	439.6125	430.6125	GB7KH	DMR/3	Kelvedon Hatch	M1GEO
DVU49	439.6125	430.6125	GB7TH	DMR/3	Broadstairs	M0LMK
DVU49	439.6125	430.6125	GB7TY		Hexham	G1HZI
DVU50	439.625	430.625	GB7EK	DMR/3	Whitstable	G6MRI
DVU50	439.625	430.625	GB7GG	DMR/1	Airdrie	GM4AUP
DVU50	439.625	430.625	GB7HB	DMR/1	Tandragee	MI0IRZ
DVU50	439.625	430.625	GB7OZ	DMR/8	Oswestry	G0DNI
DVU50	439.625	430.625	GB7RV	DMR/2	Blackburn	M0NWI
DVU51	439.6375	430.6375	GB7CL		Clacton On Sea	G0MBA
DVU51	439.6375	430.6375	GB7HM	DMR/1	Caergwrle	G1SYG
DVU51	439.6375	430.6375	GB7JF		Northampton	G7HIF
DVU51	439.6375	430.6375	GB7MH	DMR/2	East Grinstead	G3NZP
DVU51	439.6375	430.6375	GB7MJ	DMR/5	Romsey	G3ZXX
DVU51	439.6375	430.6375	GB7SR	DMR/2	Sheffield	M0GAV
DVU52	439.65	430.65	GB7DK-B		Stranraer	GM0HPK
DVU52	439.65	430.65	GB7IN	DMR/1	Alfreton	G0LCG
DVU53	439.6625	430.6625	GB7BE		Beccles	M0JGX
DVU53	439.6625	430.6625	GB7DB		Ampthill	G3YQO
DVU53	439.6625	430.6625	GB7DD	DMR/1	Dundee	MM0DUN
DVU53	439.6625	430.6625	GB7DJ	DMR/3	Northwich	M0WTX
DVU53	439.6625	430.6625	GB7FW	DMR/1	Birmingham	G8VIQ
DVU53	439.6625	430.6625	GB7KM	DMR/3	Cirencester	G0RMA
DVU53	439.6625	430.6625	GB7LE	DMR/2	Leeds	G1XCC
DVU53	439.6625	430.6625	GB7LY	DMR/1	Derry/Londonderry	GI4YWT
DVU54	439.675	430.675	GB3NS	82.5Hz	Caterham	G0OLX
DVU54	439.675	430.675	GB7AA	DMR/1	Bristol	G4CJZ
DVU54	439.675	430.675	GB7DL		Friskney	M0VBR
DVU54	439.675	430.675	GB7ED	DMR/5	Exeter	M0ZZT
DVU54	439.675	430.675	GB7SN	DMR/1	Sheffield	M1ERS
DVU55	439.6875	430.6875	GB7GF	DMR/3	Guildford	G4EML
DVU55	439.6875	430.6875	GB7NB		Norwich	G0LGJ
DVU55	439.6875	430.6875	GB7NY	DMR/1	Newry	MI0PYN
DVU55	439.6875	430.6875	GB7TP	DMR/1	Shipley	M0IRK
DVU55	439.6875	430.6875	GB7ZI		Stafford	G4YFF
DVU56	439.7	430.7	GB7AU		Amersham	G0RDI
DVU56	439.7	430.7	GB7CD		Cardiff	GW6CUR
DVU56	439.7	430.7	GB7DX		Ashford Kent	G0GCQ
DVU56	439.7	430.7	GB7MB	DMR/1	Heysham	G4TUZ
DVU56	439.7	430.7	GB7VO		Leominster	G8XYJ
DVU57	439.7125	430.7125	GB7EE	DMR/1	Edinburgh	GM7RYR
DVU57	439.7125	430.7125	GB7EL	DMR/2	Nelson	G4BLH
DVU57	439.7125	430.7125	GB7EX	DMR/3	Southend On Sea	G8YPK
DVU57	439.7125	430.7125	GB7KA	DMR/3	Kilrea	MI0AAZ
DVU57	439.7125	430.7125	GB7PT		Royston	G4NBS
DVU57	439.7125	430.7125	GB7RE	DMR/1	Retford	M0CMN
DVU57	439.7125	430.7125	GB7YS	DMR/1	Yeovil	G3ZXX
DVU58	439.725	430.725	GB7FF		Blackpool	G0WDA
DVU58	439.725	430.725	GB7ME	DMR/3	Rugby	M0IJS
DVU59	439.7375	430.7375	GB7BJ	DMR/13	Bromyard	G1MAW
DVU59	439.7375	430.7375	GB7BK	DMR/3	Reading	G8DOR
DVU59	439.7375	430.7375	GB7CK	DMR/3	Folkestone	M1CMN
DVU59	439.7375	430.7375	GB7MR	DMR/2	Oldham	G8UVC
DVU61	439.7625	430.7625	GB7AD		Bristol	G4CJZ
DVU61	439.7625	430.7625	GB7EG	DMR/2	East Grinstead	G7KBR
DVU61	439.7625	430.7625	GB7PI		Royston	M0ZPU
DVU61	439.7625	430.7625	GB7WP		Birkenhead	G4BKF
DVU62	439.775	430.775	GB7CS		East Kilbride	GM7GDE
DVU62	439.775	430.775	GB7SX	88.5Hz	Bognor Regis	G0AFN
23CM	1241.075	1241.075	GB7YD-A		Barnsley	G4LUE
23CM	1290.65	1270.65	GB7IC-A		Herne Bay	G4TKR
RM0	1297	1291	GB3NO	94.8Hz	Norwich	G8VLL
RM2	1297.05	1291.05	GB3FM	100.0Hz	Farnham	G4EPX
RM3	1297.075	1291.075	GB3PS	77.0Hz	Royston	G4NBS
RM3	1297.075	1291.075	GB3SE	103.5Hz	Stoke On Trent	G8NSS
RM6	1297.15	1291.15	GB3MM	67.0Hz	Wolverhampton	G4OKE
RM12	1297.3	1291.3	GB7YD-A		Barnsley	G4LUE
RM14A	1297.35	1291.35	GB3AK	94.8Hz	Bristol	G4CJZ
RM15	1297.375	1291.375	GB3WC	82.5Hz	Wakefield	G1XCC
RM0A	1299.85	1293.85	GB3BK	103.5Hz	Bromley	G0WYG

Internet-linked (alphabetical)

Callsign	Ch	Location	Echolink	IRLP
GB3AG	RV58	Forfar	117931	
GB3AM	R50-13	Amersham	4125	
GB3AR	RV56	Caernarfon	206003	
GB3BM	RV57	Birmingham		5702
GB3BN	RB0	Bracknell	1938	
GB3CA	RB13	Carlisle	412685	5280
GB3CG	RV58	Gloucester	190502	
GB3CH	RB02	Liskeard		5992
GB3DC	RV55	Derby	92369	
GB3DQ	RU78	Polperro	418341	5612
GB3DU	RV61	Duns	276441	
GB3DV	RB01	Maltby	120618	5130
GB3DX	RB12	Derry/Londonderry	7125	
GB3EE	RB12	Chesterfield		5046
GB3EK	RU71	Margate	48360	
GB3FH	R50-06	Somerset	228585	5361
GB3FK	RV60	Folkestone	235976	
GB3HE	RB14	Hastings	71066	
GB3HH	RV56	Buxton	97616	
GB3IE	RU68	Plymouth	27871	
GB3IK	RV61	Rochester	263025	
GB3IM-C	RU66	Douglas Iom	464453	
GB3IM-R	RU66	Ramsey Iom	464453	
GB3IM-S	RB05	Douglas Iom	464453	
GB3IN	RV51	Alfreton	98258	
GB3IR	RV61	Richmond Yorks	1353	5562
GB3IW	RB09	Ryde	401932	
GB3IW		East Cowes	401932	
GB3JS	RB01	Great Yarmouth	246617	
GB3KC	RU72	Stourbridge	430900	
GB3KD	RV63	Kidderminster	78750	
GB3KE	RV55	Glasgow	5411	5410
GB3KL	RB04	Kings Lynn	77266	
GB3KR	RB03	Kidderminster	4304	
GB3KS	RV50	Dover	346463	
GB3KU	RB03	Ashton-Under-Lyne	1234567	
GB3LF	RB14	Kendal	184457	5140
GB3LR	RU69	Newhaven	494669	
GB3LS	RB02	Lincoln	268511	
GB3LV	RB02	North London	155403	5600
GB3MH	RV50	East Grinstead	453929	5569
GB3MI	RV57	Manchester	197681	
GB3NC	RV58	Roche	282184	
GB3ND	RB14	Bideford	221334	
GB3NK	RB04	Erith	54760	
GB3NU	RU71	Sheringham	388653	
GB3OA	RV49	Southport	5302	5302
GB3PA	RV50	Paisley	116678	
GB3PY	RB08	Cambridge	222303	
GB3PZ	RU72	Dukinfield	2591	5400
GB3SB	RV52	Selkirk	116678	
GB3SD	RB14	Weymouth	112689	
GB3TD	RB03	Swindon	43307	
GB3TR	RV52	Torquay		5582
GB3UB	RB04	Bath Avon	201135	
GB3XN	RU74	Worksop	153126	5708
GB3YL		Lowestoft	227697	
GB3ZB	RU66	Bristol		5429
GB7SJ	RB07	Northwich	455339	41360
MB7ADE		Kenilworth	176074	
MB7AJS		Burntwood		5269
MB7APR		Folkestone	24	

Digital Nodes including D-Star (by Channel)

Ch No	Callsign	Location	Type
DVU13	GB7BS	Bristol	DMR
DVU13	GB7FU	Pointon	DMR
DVU13	GB7NS	Caterham	DMR
DVU13	GB7TD	Wakefield	DMR
DVU32	GB7AL	Ipswich	DMR
DVU32	GB7BM	Birmingham	D-STAR
DVU32	GB7CW	Bridgend	DMR
DVU32	GB7EP	Epsom	DMR
DVU32	GB7LN	Lincoln	DMR
DVU32	GB7LP	Liverpool	DMR
DVU33	GB7DN	Dungiven	D-STAR
DVU33	GB7GT	Ludlow	DMR
DVU33	GB7JD	Jedburgh	DMR/DSTAR
DVU33	GB7ND	Great Ellingham	DMR
DVU33	GB7SD	Weymouth	DMR
DVU33	GB7WI	Walkington	DMR
DVU34	GB7AS	Ashford	DMR
DVU34	GB7DR	Poole	DMR
DVU34	GB7DS	Norwich	ANL/DMR
DVU34	GB7GB	Birmingham	DMR
DVU34	GB7HS	Cleckheaton	DMR
DVU34	GB7KB	Welham Green	D-STAR
DVU34	GB7PN	Prestatyn	DMR
DVU35	GB7AV	Aylesbury	DMR
DVU35	GB7BX	Much Wenlock	DMR
DVU35	GB7ES	Eastbourne	D-STAR
DVU35	GB7FO	Blackpool	DMR
DVU35	GB7GC	Grimsby	DMR/DSTAR
DVU35	GB7PP	Ipswich	D-STAR
DVU36	GB7BP	Milton Keynes	D-STAR
DVU36	GB7HR	Heathrow	DMR
DVU36	GB7IC-B	Herne Bay	D-STAR
DVU36	GB7NE	Ashington	D-STAR
DVU36	GB7SU	Southampton	DMR
DVU36	GB7TT	Troon	DMR
DVU37	GB7JB	Mere Wiltshire	DMR
DVU37	GB7NL	Exeter	D-STAR
DVU37	GB7SK	Leicester	DMR
DVU37	GB7WL	Amersham	DMR
DVU37	GB7XX	Felling	DMR
DVU38	GB7CH	Louth	D-STAR
DVU38	GB7LR	Leicester	DMR
DVU38	GB7SC	Bognor Regis	DMR
DVU38	GB7SE	Thurrock	DMR
DVU38	GB7TV	New Marske	D-STAR
DVU38	GB7YZ	Mold	MULTI
DVU39	GB7DC	Derby	MULTI
DVU39	GB7HU	South Cave	D-STAR
DVU39	GB7MW	Carrickfergus	MULTI
DVU39	GB7NF	Newhaven	DMR
DVU39	GB7WB	Weston-Super-Mare	MULTI
DVU39	GB7WC	Warrington	D-STAR
DVU39	GB7WF	Bewdley	D-STAR
DVU39	GB7ZP	Chelmsford	D-STAR
DVU40	GB7DP-B	Dundee	D-STAR
DVU40	GB7FR	Worthing	DMR
DVU40	GB7KT	Andover	DMR
DVU40	GB7PE	Peterborough	DMR
DVU40	GB7UZ	Lancaster	DMR
DVU41	GB7AH	Ahoghill	MULTI
DVU41	GB7CF	Folkestone	FUSION
DVU41	GB7IT	Weston-Super-Mare	DMR
DVU41	GB7LO	Bromley	DMR
DVU41	GB7NO	Northampton	DMR
DVU41	GB7YD-B	Barnsley	D-STAR
DVU42	GB7AK	Barking	DMR
DVU42	GB7PK	Portsmouth	DMR
DVU42	GB7SI	Stoke-On-Trent	DMR
DVU42	GB7TC	Swindon	DMR
DVU42	GB7UL	Carrickfergus	DMR
DVU42	GB7YR	Doncaster	DMR
DVU43	GB7DG	Bromsgrove	D-STAR
DVU43	GB7EN	Wellingborough	MULTI
DVU43	GB7HE	Hastings	D-STAR
DVU43	GB7IK	Rochester	DMR
DVU43	GB7IS	Weston-Super-Mare	C4FM
DVU43	GB7LF	Lancaster	D-STAR
DVU43	GB7MC	St Austell	D-STAR
DVU43	GB7NU	Sheringham	D-STAR
DVU46	GB7EB	Beccles	DMR
DVU46	GB7HA	Halstead	DMR
DVU46	GB7HX	Huddersfield	DMR
DVU48	GB7MK	Ipswich	DMR
DVU48	GB7RR	Nottingham	DMR
DVU48	GB7WT	Omagh	D-STAR
DVU49	GB7DV	St. Helens	D-STAR
DVU49	GB7FH	Fareham	D-STAR
DVU49	GB7IQ	Stafford	DMR
DVU49	GB7KH	Kelvedon Hatch	MULTI
DVU49	GB7TH	Broadstairs	DMR
DVU49	GB7TY	Hexham	D-STAR
DVU50	GB7EK	Whitstable	DMR
DVU50	GB7GG	Airdrie	DMR
DVU50	GB7HB	Tandragee	DMR
DVU50	GB7OZ	Oswestry	DMR
DVU50	GB7RV	Blackburn	DMR
DVU51	GB7CL	Clacton On Sea	DMR
DVU51	GB7HM	Caergwrle	DMR
DVU51	GB7JF	Northampton	D-STAR
DVU51	GB7MH	East Grinstead	DMR/DSTAR
DVU51	GB7MJ	Romsey	DMR
DVU51	GB7SR	Sheffield	DMR
DVU52	GB7DK-B	Stranraer	D-STAR
DVU52	GB7IN	Alfreton	DMR
DVU53	GB7BE	Beccles	D-STAR
DVU53	GB7DB	Ampthill	D-STAR
DVU53	GB7DD	Dundee	DMR
DVU53	GB7DJ	Northwich	DMR
DVU53	GB7FW	Birmingham	DMR
DVU53	GB7KM	Cirencester	DMR
DVU53	GB7LE	Leeds	DMR
DVU53	GB7LY	Derry/Londonderry	DMR
DVU54	GB7AA	Bristol	ANL/DMR
DVU54	GB7DL	Friskney	D-STAR
DVU54	GB7ED	Exeter	DMR
DVU54	GB7SN	Sheffield	DMR
DVU55	GB7GF	Guildford	DMR
DVU55	GB7NB	Norwich	D-STAR
DVU55	GB7NY	Newry	DMR
DVU55	GB7TP	Shipley	MULTI
DVU55	GB7ZI	Stafford	D-STAR
DVU56	GB7AU	Amersham	D-STAR
DVU56	GB7CD	Cardiff	D-STAR
DVU56	GB7DX	Ashford Kent	D-STAR
DVU56	GB7MB	Heysham	DMR
DVU56	GB7VO	Leominster	D-STAR
DVU57	GB7EE	Edinburgh	DMR
DVU57	GB7EL	Nelson	DMR
DVU57	GB7EX	Southend On Sea	DMR
DVU57	GB7KA	Kilrea	MULTI
DVU57	GB7PT	Royston	MULTI
DVU57	GB7RE	Retford	DMR
DVU57	GB7YS	Yeovil	DMR
DVU58	GB7FF	Blackpool	C4FM
DVU58	GB7ME	Rugby	DMR
DVU59	GB7BJ	Bromyard	DMR
DVU59	GB7BK	Reading	DMR
DVU59	GB7CK	Folkestone	DMR
DVU59	GB7MR	Oldham	DMR
DVU61	GB7AD	Bristol	D-STAR
DVU61	GB7EG	East Grinstead	DMR
DVU61	GB7PI	Royston	D-STAR
DVU61	GB7WP	Birkenhead	D-STAR
DVU62	GB7CS	East Kilbride	C4FM
DVU62	GB7SX	Bognor Regis	C4FM
RB0	GB3DT	Blandford	FUSION
RB0	GB3MK	Milton Keynes	FUSION
RB0	GB3NT	Newcastle Upon Tyne	FUSION
RB0	GB3PU	Perth	FUSION
RB04	GB3SP	Pembroke	FUSION
RB04	GB7HF	Welham Green	FUSION
RB05	GB7PD-B	Pembroke Dock	D-STAR
RB06	GB3HC	Hereford	FUSION
RB07	GB3MS	Worcester	FUSION
RB07	GB3TS	Sunderland	FUSION
RB07	GB7SJ	Northwich	FUSION
RB08	GB3AN	Amlwch	FUSION
RB08	GB3TF	Telford	FUSION
RB09	GB3HD	Huddersfield	FUSION
RB10	GB3DD	Dundee	FUSION
RB12	GB3TJ	Corbridge	FUSION
RB14	GB3WF	Otley	FUSION
RB15	GB3HB	Roche	FUSION
RB15	GB3LH	Shrewsbury	FUSION
RB15	GB3TH	Tamworth	FUSION
RB15	GB3WI	Wisbech	FUSION
RM12	GB7YD-A	Barnsley	D-STAR
RU66	GB3KW	Glasgow	FUSION
RU66	GB3UO	Chirk	FUSION
RU68	GB3GR	Grantham	FUSION
RU69	GB7DO	Doncaster	ANL/DMR
RU70	GB7WX	Tarvin	FUSION
RU71	GB3BP	Belper	FUSION
RU71	GB3NP	Towcester	FUSION
RU71	GB3XL	Shipley	ANL/DMR
RU71	GB7FI	Axbridge	ANL/DMR
RU71	GB7JM	Louth	DMR
RU71	GB7TQ	Torquay	FUSION
RU72	GB3GL	Glasgow	FUSION
RU72	GB3JC	Norwich	FUSION
RU73	GB3XX	Daventry	FUSION
RU74	GB3DM	Dumbarton	FUSION
RU74	GB3FI	Cheddar	ANL/DMR
RU74	GB3VN	Ludlow	FUSION
RU74	GB7BR	Ramsey Iom	DMR
RU74	GB7CA	Douglas	DMR
RU76	GB3OM	Omagh	ANL/DMR
RU76	GB7BD	Bristol	DMR
RU76	GB7FG	Carmarthen	FUSION
RU76	GB7RY	Rye	FUSION
RU77	GB7JL	Wigan	ANL/DMR
RU77	GB7SH	Sheffield	FUSION
RU78	GB3ZI	Stafford	FUSION
RU78	GB7FC	Blackpool	MULTI
RU78	GB7KE	Kettering	FUSION
RV48	GB3CF	Markfield	FUSION
RV48	GB3SS	Elgin Moray	FUSION
RV48	GB3WR	Cheddar	FUSION
RV48	GB7MA	Bury	D-STAR
RV48	GB7RW	Whitby	ANL/DSTAR
RV48	GB7SB	Brighton	D-STAR
RV49	GB3NA	Barnsley	FUSION
RV49	GB3TO	Northampton	ANL/DSTAR
RV49	GB3VM	Ludlow	FUSION
RV50	GB3NW	Worcester	FUSION
RV51	GB3IN	Alfreton	MULTI
RV51	GB3VO	Chirk	FUSION
RV51	GB7CT	Tring	DMR
RV51	GB7DE	Edinburgh	MULTI
RV51	GB7RN	Fareham	D-STAR
RV52	GB3HS	Walkington	FUSION
RV52	GB3MN	Disley	FUSION
RV53	GB3AA	Bristol	FUSION
RV53	GB3DW	Harlech	FUSION
RV53	GB3KI	Herne Bay	ANL/DSTAR
RV53	GB7IC-C	Herne Bay	ANL/DSTAR
RV53	GB7PB	Durham	D-STAR
RV54	GB3PR	Perth	FUSION
RV54	GB7IE	Plymouth	C4FM
RV54	GB7RB-C	Cowbridge	D-STAR
RV54	GB7YD-C	Barnsley	D-STAR
RV55	GB3CD	Crook	FUSION
RV55	GB3KE	Glasgow	FUSION
RV55	GB3WE	Weston-Super-Mare	DMR/DSTAR
RV55	GB3XP	Morden	FUSION
RV55	GB7DK	Stranraer	FUSION
RV55	GB7GD	Aberdeen	D-STAR
RV56	GB7CM	Blandford	FUSION
RV57	GB3FG	Carmarthen	FUSION
RV57	GB3MI	Manchester	ANL/DSTAR
RV57	GB7OK	Bromley	D-STAR
RV58	GB3LM	Lincoln	FUSION
RV58	GB3NC	Roche	FUSION
RV58	GB3TW	Gateshead	FUSION
RV58	GB3VT	Stoke On Trent	ANL/DSTAR
RV59	GB3ZA	Hereford	FUSION
RV59	GB7PD-C	Pembroke Dock	D-STAR
RV59	GB7SF	Sheffield	D-STAR
RV60	GB3HA	Corbridge	FUSION
RV60	GB7NI	Carrickfergus	MULTI
RV61	GB3IP	Stafford	DMR
RV61	GB7IP	Leicester	ANL/DMR
RV62	GB3DG	Newton Stewart	FUSION
RV62	GB3RF	Accrington	MULTI
RV62	GB7DA	Airdrie	D-STAR
RV62	GB7ER	Exeter	FUSION
RV62	GB7IV	Chandlers Ford	D-STAR
RV62	GB7TE	Clacton On Sea	D-STAR
RV63	GB3GO	Llandudno	C4FM
RV63	GB7FK-C	Folkestone	D-STAR
	GB7IC-A	Herne Bay	D-STAR
	GB7ZZ	Herne Bay	D-STAR

Locators

The IARU Locator System, usually just called 'Locator', provides a means of pinpointing stations throughout the world. It is most often used by operators above 30MHz, as a means of calculating the distance between two stations. It is also used on the 136kHz band for the same reason. For use by operators on the upper microwave bands, it can have eight digits, though only the first six are dealt with here. The system is based upon latitude and longitude.

As the map and diagrams show, there are three sizes of 'rectangle'. The largest, known as a 'field', is 20° of longitude (east-west) by 10° latitude (north-south), and is designated by two letters. Most of Britain is in IO field. The next rectangle, known as a 'square' (though it is actually neither truly square nor rectangular!) is 2° of longitude by 1° of latitude. One hundred squares make up one field and, as the map shows, these are given numbers 00 in the south-west corner to 99 in the north-east. Dublin is in IO63. Finally, each square is divided into 576 'sub-squares', 5 minutes of longitude by 2.5 minutes of latitude, and given letters from AA to XX.

To find out your locator, first use a map of your area to determine your exact latitude and longitude, then use the map on this page and the squares diagram opposite to pinpoint your locator. Computer programs and online calculators are available to do this more easily, especially for those who operate from various locations.

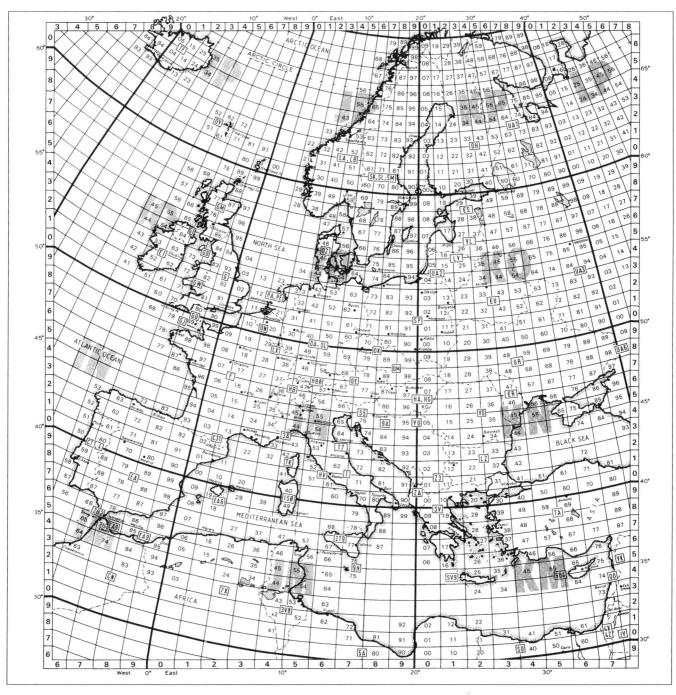

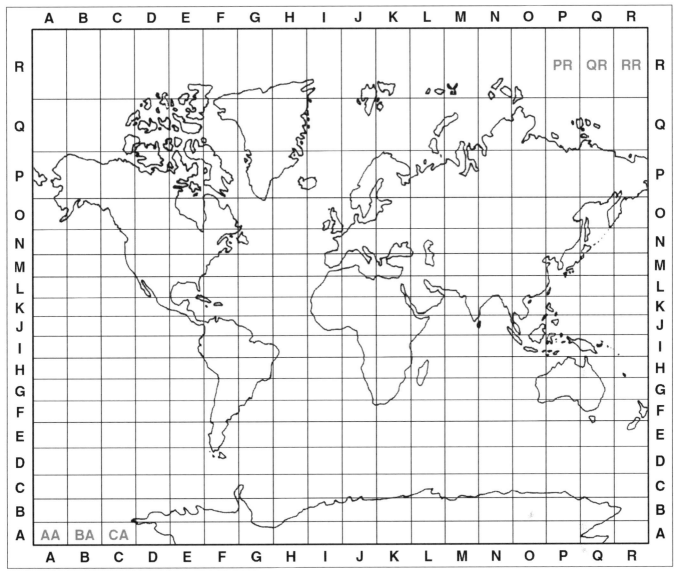

The IARU Locator system may be used throughout the world without repeats. The map above shows the fields that make up the first two letters of the Locator. Examples are shown at two of the corners. The map left shows numbering of squares within the fields.

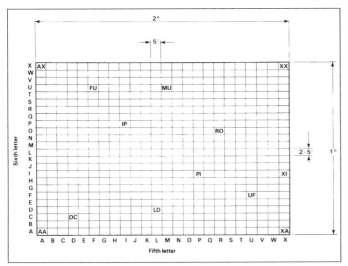

A square (the numbered part of the Locator) is divided into 576 sub-squares, designated AA to XX. Each sub-square is 5' W-E and 2.5' N-S.

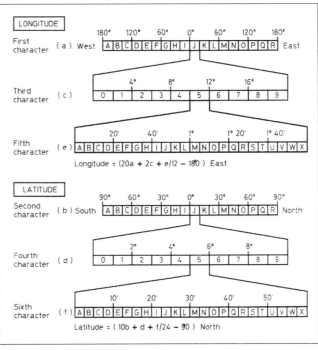

Longitude = (20a + 2c + e/12 − 180) East

Latitude = (10b + d + f/24 − 90) North

The final two letters may be calculated thus.

Prefix List

Callsigns for the world's nations are determined by the International Telecommunications Union (ITU). This is the United Nations agency that co-ordinates radio activity for all spectrum users. The prefixes used by a country for both commercial and amateur radio purposes are determined from one or more ITU allocation blocks issued to that country. The amateur radio callsigns in use for a particular country might use one or a number of combinations derived from the authorised ITU allocation(s) for that country. The following list shows callsign prefixes currently in use. Most are derived from the callsign blocks allocated to administrations by the ITU for use within the countries, territories and dependencies for which a country is responsible. Also shown are some unauthorised prefixes which may be heard and which may or may not be recognised as a DXCC entity, eg 1A0 (SMOM). 1B (the Turkish area of North Cyprus) and 1Z (Karea State - Myanmar) are unofficial and are not recognised for DXCC purposes, so these are not shown.

Full information on prefixes is contained in the *RSGB Prefix Guide*.

Prefix	Entity	Cont.	ITU	CQ	1.8	3.5	7.0	10.1	14	18	21	24	28	50	144	oth.
1A	Sov. Mil. Order of Malta	EU	28	15												
3A	Monaco	EU	27	14												
3B6, 7	Agalega & St. Brandon Is.	AF	53	39												
3B8	Mauritius	AF	53	39												
3B9	Rodriguez I.	AF	53	39												
3C	Equatorial Guinea	AF	47	36												
3C0	Annobon I.	AF	52	36												
3D2	Fiji	OC	56	32												
3D2	Conway Reef	OC	56	32												
3D2	Rotuma I.	OC	56	32												
3DA	Swaziland	AF	57	38												
3V	Tunisia	AF	37	33												
3W, XV	Vietnam	AS	49	26												
3X	Guinea	AF	46	35												
3Y	Bouvet	AF	67	38												
3Y	Peter 1 I.	AN	72	12												
4J, 4K	Azerbaijan	AS	29	21												
4L	Georgia	AS	29	21												
4O	Montenegro	EU	28	15												
4S	Sri Lanka	AS	41	22												
4U_ITU	ITU HQ	EU	28	14												
4U_UN	United Nations HQ	NA	08	05												
4W	Timor - Leste	OC	54	28												
4X, 4Z	Israel	AS	39	20												
5A	Libya	AF	38	34												
5B, C4, P3	Cyprus	AS	39	20												
5H-5I	Tanzania	AF	53	37												
5N	Nigeria	AF	46	35												
5R	Madagascar	AF	53	39												
5T	Mauritania	AF	46	35												
5U	Niger	AF	46	35												
5V	Togo	AF	46	35												
5W	Samoa	OC	62	32												
5X	Uganda	AF	48	37												
5Y-5Z	Kenya	AF	48	37												
6V-6W	Senegal	AF	46	35												
6Y	Jamaica	NA	11	08												
7O5	Yemen	AS	39	21												
7P	Lesotho	AF	57	38												
7Q	Malawi	AF	53	37												
7T-7Y	Algeria	AF	37	33												
8P	Barbados	NA	11	08												
8Q	Maldives	AS/AF	41	22												
8R	Guyana	SA	12	09												
9A	Croatia	EU	28	15												
9G	Ghana	AF	46	35												
9H	Malta	EU	28	15												
9I-9J	Zambia	AF	53	36												
9K	Kuwait	AS	39	21												
9L	Sierra Leone	AF	46	35												
9M2, 4	West Malaysia	AS	54	28												
9M6, 8	East Malaysia	OC	54	28												
9N	Nepal	AS	42	22												
9Q-9T	Dem. Rep. of Congo	AF	52	36												

Prefix	Entity	Cont.	ITU	CQ	1.8	3.5	7.0	10.1	14	18	21	24	28	50	144	oth.
9U	Burundi	AF	52	36												
9V	Singapore	AS	54	28												
9X	Rwanda	AF	52	36												
9Y-9Z	Trinidad & Tobago	SA	11	09												
A2	Botswana	AF	57	38												
A3	Tonga	OC	62	32												
A4	Oman	AS	39	21												
A5	Bhutan	AS	41	22												
A6	United Arab Emirates	AS	39	21												
A7	Qatar	AS	39	21												
A9	Bahrain	AS	39	21												
AP	Pakistan	AS	41	21												
B	China	AS	33, 42-44	23, 24												
BS7	Scarborough Reef	AS	50	27												
BU-BX	Taiwan	AS	44	24												
BV9P	Pratas I.	AS	44	24												
C2	Nauru	OC	65	31												
C3	Andorra	EU	27	14												
C5	The Gambia	AF	46	35												
C6	Bahamas	NA	11	08												
C8-9	Mozambique	AF	53	37												
CA-CE	Chile	SA	14,16	12												
CE0	Easter I.	SA	63	12												
CE0	Juan Fernandez Is.	SA	14	12												
CE0	San Felix & San Ambrosio	SA	14	12												
CE9/KC4	Antarctica	AN	67,69-74	S												
CM, CO	Cuba	NA	11	08												
CN	Morocco	AF	37	33												
CP	Bolivia	SA	12,14	10												
CT	Portugal	EU	37	14												
CT3	Madeira Is.	AF	36	33												
CU	Azores	EU	36	14												
CV-CX	Uruguay	SA	14	13												
CY0	Sable I.	NA	09	05												
CY9	St. Paul I.	NA	09	05												
D2-3	Angola	AF	52	36												
D4	Cape Verde	AF	46	35												
D6	Comoros	AF	53	39												
DA-DR	Fed. Rep. of Germany	EU	28	14												
DU-DZ	Philippines	OC	50	27												
E3	Eritrea	AF	48	37												
E4	Palestine	AS	39	20												
E5	N. Cook Is.	OC	62	32												
E5	S. Cook Is.	OC	62	32												
E6	Niue	OC	62	32												
E7	Bosnia-Herzegovina	EU	28	15												
EA-EH	Spain	EU	37	14												
EA6-EH6	Balearic Is.	EU	37	14												
EA8-EH8	Canary Is.	AF	36	33												
EA9-EH9	Ceuta & Melilla	AF	37	33												
EI-EJ	Ireland	EU	27	14												
EK	Armenia	AS	29	21												
EL	Liberia	AF	46	35												
EP-EQ	Iran	AS	40	21												
ER	Moldova	EU	29	16												
ES	Estonia	EU	29	15												
ET	Ethiopia	AF	48	37												
EU-EW	Belarus	EU	29	16												
EX	Kyrgyzstan	AS	30, 31	17												
EY	Tajikistan	AS	30	17												
EZ	Turkmenistan	AS	30	17												
F	France	EU	27	14												
FG, TO	Guadeloupe	NA	11	08												
FH, TO	Mayotte	AF	53	39												
FJ, TO	Saint Barthelemy	NA	11	08												
FK, TX	New Caledonia	OC	56	32												
FK, TX	Chesterfield Is.	OC	56	30												
FM, TO	Martinique	NA	11	08												
FO, TX	Austral I.	OC	63	32												
FO, TX	Clipperton I.	NA	10	07												
FO, TX	French Polynesia	OC	63	32												
FO, TX	Marquesas Is.	OC	63	31												

Prefix	Entity	Cont.	ITU	CQ	1.8	3.5	7.0	10.1	14	18	21	24	28	50	144	oth.
FP	St. Pierre & Miquelon	NA	09	05												
FR, TO	Reunion I.	AF	53	39												
FT/G, TO	Glorioso Is.	AF	53	39												
FT/J,E, TO	Juan de Nova, Europa	AF	53	39												
FT/T, TO	Tromelin I.	AF	53	39												
FS, TO	Saint Martin	NA	11	08												
FT/W	Crozet I.	AF	68	39												
FT/X	Kerguelen Is.	AF	68	39												
FT/Z	Amsterdam & St. Paul Is.	AF	68	39												
FW	Wallis & Futuna Is.	OC	62	32												
FY	French Guiana	SA	12	09												
G, GX, M, MX, 2E	England	EU	27	14												
GD, GT, MD, MT, 2D	Isle of Man	EU	27	14												
GI, GN, MI, MN, 2I	Northern Ireland	EU	27	14												
GJ, GH, MJ, MH, 2J	Jersey	EU	27	14												
GM, GS, MM, MS, 2M	Scotland	EU	27	14												
GU, GP, MU, MP, 2U	Guernsey	EU	27	14												
GW, GC, MW, MC, 2W	Wales	EU	27	14												
H4	Solomon Is.	OC	51	28												
H40	Temotu Province	OC	51	32												
HA, HG	Hungary	EU	28	15												
HB	Switzerland	EU	28	14												
HB0	Liechtenstein	EU	28	14												
HC-HD	Ecuador	SA	12	10												
HC8-HD8	Galapagos Is.	SA	12	10												
HH	Haiti	NA	11	08												
HI	Dominican Republic	NA	11	08												
HJ-HK, 5J-5K	Colombia	SA	12	09												
HK0	Malpelo I.	SA	12	09												
HK0	San Andres & Providencia	NA	11	07												
HL, 6K-6N	Republic of Korea	AS	44	25												
HO-HP	Panama	NA	11	07												
HQ-HR	Honduras	NA	11	07												
HS, E2	Thailand	AS	49	26												
HV	Vatican	EU	28	15												
HZ	Saudi Arabia	AS	39	21												
I	Italy	EU	28	15,33												
IS0, IM0	Sardinia	EU	28	15												
J2	Djibouti	AF	48	37												
J3	Grenada	NA	11	08												
J5	Guinea-Bissau	AF	46	35												
J6	St. Lucia	NA	11	08												
J7	Dominica	NA	11	08												
J8	St. Vincent	NA	11	08												
JA-JS, 7J-7N	Japan	AS	45	25												
JD	Minami Torishima	OC	90	27												
JD	Ogasawara	AS	45	27												
JT-JV	Mongolia	AS	32,33	23												
JW	Svalbard	EU	18	40												
JX	Jan Mayen	EU	18	40												
JY	Jordan	AS	39	20												
K, W, N, AA-AK	United States of America	NA	6,7,8	3,4,5												
KG4	Guantanamo Bay	NA	11	08												
KH0	Mariana Is.	OC	64	27												
KH1	Baker & Howland Is.	OC	61	31												
KH2	Guam	OC	64	27												
KH3	Johnston I.	OC	61	31												
KH4	Midway I.	OC	61	31												
KH5	Palmyra & Jarvis Is.	OC	61, 62	31												
KH5K	Kingman Reef	OC	61	31												
KH6,7	Hawaii	OC	61	31												
KH7K	Kure I.	OC	61	31												
KH8	American Samoa	OC	62	32												
KH8	Swains I.	OC	62	32												
KH9	Wake I.	OC	65	31												
KL, AL, NL, WL	Alaska	NA	1, 2	1												
KP1	Navassa I.	NA	11	08												
KP2	Virgin Is.	NA	11	08												
KP3, 4	Puerto Rico	NA	11	08												
KP5	Desecheo I.	NA	11	08												
LA-LN	Norway	EU	18	14												
LO-LW	Argentina	SA	14,16	13												

Prefix	Entity	Cont.	ITU	CQ	1.8	3.5	7.0	10.1	14	18	21	24	28	50	144	oth.
LX	Luxembourg	EU	27	14												
LY	Lithuania	EU	29	15												
LZ	Bulgaria	EU	28	20												
OA-OC	Peru	SA	12	10												
OD	Lebanon	AS	39	20												
OE	Austria	EU	28	15												
OF-OI	Finland	EU	18	15												
OH0	Aland Is.	EU	18	15												
OJ0	Market Reef	EU	18	15												
OK-OL	Czech Republic	EU	28	15												
OM	Slovak Republic	EU	28	15												
ON-OT	Belgium	EU	27	14												
OU-OW, OZ	Denmark	EU	18	14												
OX	Greenland	NA	5, 75	40												
OY	Faroe Is.	EU	18	14												
P2	Papua New Guinea	OC	51	28												
P4	Aruba	SA	11	09												
P5	DPR of Korea	AS	44	25												
PA-PI	Netherlands	EU	27	14												
PJ2	Curacao	SA	11	09												
PJ4	Bonaire	SA	11	09												
PJ5, 6	Saba & St. Eustatius	NA	11	08												
PJ7	St Maarten	NA	11	08												
PP-PY, ZV-ZZ	Brazil	SA	12, 13, 15	11												
PP0-PY0F	Fernando de Noronha	SA	13	11												
PP0-PY0S	St. Peter & St. Paul Rocks	SA	13	11												
PP0-PY0T	Trindade & Martim Vaz Is.	SA	15	11												
PZ	Suriname	SA	12	09												
R1/F	Franz Josef Land	EU	75	40												
S0	Western Sahara	AF	46	33												
S2	Bangladesh	AS	41	22												
S5	Slovenia	EU	28	15												
S7	Seychelles	AF	53	39												
S9	Sao Tome & Principe	AF	47	36												
SA-SM, 7S-8S	Sweden	EU	18	14												
SN-SR	Poland	EU	28	15												
ST	Sudan	AF	47, 48	34												
SU	Egypt	AF	38	34												
SV-SZ, J4	Greece	EU	28	20												
SV/A	Mount Athos	EU	28	20												
SV5, J45	Dodecanese	EU	28	20												
SV9, J49	Crete	EU	28	20												
T2	Tuvalu	OC	65	31												
T30	W. Kiribati (Gilbert Is.)	OC	65	31												
T31	C. Kiribati (British Phoenix Is)	OC	62	31												
T32	E. Kiribati (Line Is.)	OC	61, 63	31												
T33	Banaba I. (Ocean I.)	OC	65	31												
T5, 6O	Somalia	AF	48	37												
T7	San Marino	EU	28	15												
T8	Palau	OC	64	27												
TA-TC	Turkey	EU/AS	39	20												
TF	Iceland	EU	17	40												
TG, TD	Guatemala	NA	12	07												
TI, TE	Costa Rica	NA	11	07												
TI9	Cocos I.	NA	12	07												
TJ	Cameroon	AF	47	36												
TK	Corsica	EU	28	15												
TL	Central Africa	AF	47	36												
TN	Congo (Republic of the)	AF	52	36												
TR	Gabon	AF	52	36												
TT	Chad	AF	47	36												
TU	Cote d'Ivoire	AF	46	35												
TY	Benin	AF	46	35												
TZ	Mali	AF	46	35												
UA-UI1-7, RA-RZ	European Russia	EU	S	16												
UA2, RA2	Kaliningrad	EU	29	15												
UA-UI8, 9, 0, RA-RZ	Asiatic Russia	AS	S	S												
UJ-UM	Uzbekistan	AS	30	17												
UN-UQ	Kazakhstan	AS	29-31	17												
UR-UZ, EM-EO	Ukraine	EU	29	16												
V2	Antigua & Barbuda	NA	11	08												
V3	Belize	NA	11	07												

Prefix	Entity	Cont.	ITU	CQ	1.8	3.5	7.0	10.1	14	18	21	24	28	50	144	oth.
V4	St. Kitts & Nevis	NA	11	08												
V5	Namibia	AF	57	38												
V6	Micronesia	OC	65	27												
V7	Marshall Is.	OC	65	31												
V8	Brunei Darussalam	OC	54	28												
VA-VG, VO,VY	Canada	NA	2-4, 9, 75	1-5												
VK, AX	Australia	OC	55, 58, 59	29, 30												
VK0	Heard I.	AF	68	39												
VK0	Macquarie I.	OC	60	30												
VK9C	Cocos (Keeling) Is.	OC	54	29												
VK9L	Lord Howe I.	OC	60	30												
VK9M	Mellish Reef	OC	56	30												
VK9N	Norfolk I.	OC	60	32												
VK9W	Willis I.	OC	55	30												
VK9X	Christmas I.	OC	54	29												
VP2E	Anguilla	NA	11	08												
VP2M	Montserrat	NA	11	08												
VP2V	British Virgin Is.	NA	11	08												
VP5	Turks & Caicos Is.	NA	11	08												
VP6	Pitcairn I.	OC	63	32												
VP646	Ducie I.	OC	63	32												
VP8	Falkland Is.	SA	16	13												
VP8, LU	South Georgia I.	SA	73	13												
VP8, LU	South Orkney Is.	SA	73	13												
VP8, LU	South Sandwich Is.	SA	73	13												
VP8, LU, CE9, HF0, 4K1	South Shetland Is.	SA	73	13												
VP9	Bermuda	NA	11	05												
VQ9	Chagos Is.	AF	41	39												
VR	Hong Kong	AS	44	24												
VU	India	AS	41	22												
VU4	Andaman & Nicobar Is.	AS	49	26												
VU7	Lakshadweep Is.	AS	41	22												
XA-XI	Mexico	NA	10	06												
XA4-XI4	Revillagigedo	NA	10	06												
XT	Burkina Faso	AF	46	35												
XU	Cambodia	AS	49	26												
XW	Laos	AS	49	26												
XX9	Macao	AS	44	24												
XY-XZ	Myanmar	AS	49	26												
YA, T6	Afghanistan	AS	40	21												
YB-YH	Indonesia	OC	51,54	28												
YI	Iraq	AS	39	21												
YJ	Vanuatu	OC	56	32												
YK	Syria	AS	39	20												
YL	Latvia	EU	29	15												
YN,H6-7,HT	Nicaragua	NA	11	07												
YO-YR	Romania	EU	28	20												
YS, HU	El Salvador	NA	11	07												
YT-YU	Serbia	EU	28	15												
YV-YY, 4M	Venezuela	SA	12	09												
YV0	Aves I.	NA	11	08												
Z2	Zimbabwe	AF	53	38												
Z3	Macedonia	EU	28	15												
Z8	South Sudan (Rep of)	AF	48	34												
ZA	Albania	EU	28	15												
ZB2	Gibraltar	EU	37	14												
ZC4	UK Sov. Base Areas on Cyprus	AS		3 9												
20																
ZD7	St. Helena	AF	66	36												
ZD8	Ascension I.	AF	66	36												
ZD9	Tristan da Cunha & Gough I.	AF	66	38												
ZF	Cayman Is.	NA	11	08												
ZK3	Tokelau Is.	OC	62	31												
ZL-ZM	New Zealand	OC	60	32												
ZL7	Chatham Is.	OC	60	32												
ZL8	Kermadec Is.	OC	60	32												
ZL9	Auckland & Campbell Is.	OC	60	32												
ZP	Paraguay	SA	14	11												
ZR-ZU	South Africa	AF	57	38												
ZS8	Prince Edward & Marion Is.	AF	57	38												

Key to abbreviations of continents:
AF = Africa, **AN** = Antarctica, **AS** = Asia, **EU** = Europe, **NA** = North America, **OC** = Oceania, **SA** = South America

RSGB QSL Bureau

Whilst sending cards for a much-prized contact will always be quicker by direct mail, QSLing via the RSGB Bureau remains an extremely cost effective option, indeed the RSGB QSL Bureau enables members to exchange cards worldwide in the cheapest practical way.

How it works

QSL cards arriving at the central bureau are initially separated into UK and Foreign destinations. Overseas cards are sent in bulk to other member societies of the International Amateur Radio Union (IARU). Cards for stations within the UK are sorted into separate callsign groups and sent to the appropriate volunteer collection managers, on a quarterly schedule. They place cards in stamped addressed envelopes (SAEs) provided to them by the call holders.

Who can use The Bureau?

Unlike the RSGB, many other national societies make extra charges for using their QSL service. The RSGB QSL Bureau is an inclusive membership service and operates as follows:

- **UK RSGB members** can send and receive their personal cards without additional charges, subject to the conditions shown here.
- **UK non-RSGB members** can collect their personal cards only by using the '*Pay-to-Receive*' service but cannot send cards via the bureau. See RSGB website for details
- **Overseas RSGB members** can send their outgoing cards to the RSGB QSL bureau for distribution. UK call holders should collect in the normal way, via their UK call-sign. Non-UK call holders should arrange collection via a UK-based QSL manager, who should also be a member.
- **Overseas non-RSGB members** may send cards addressed to UK-based stations only.
- **Affiliated Societies and independent QSL Managers** can send their own cards and those for club members or stations for whom they act, but should include current RSGB membership details for every station whose cards they wish to send. Cards included from overseas stations and intended for delivery outside the UK will not processed without proof of membership and will not be returned.

Available Destinations

A full list of IARU partner QSL bureaus can be found at: *www.iaru.org/iaruqsl.html* Keeping an up-to-date copy to hand is vital when deciding which route to send your card. For example, there are currently no bureaus in, Egypt, Kazakhstan, Morocco, and Mauritius, Sudan and several other African and Caribbean countries, plus many more smaller destinations.

Activity also relates to the frequency with which cards can be dispatched to a particular destination. This may range from monthly to annually, according to demand and is something to consider before sending your card via the bureau.

Responsible QSLing

The Bureau handles approximately 1.5 million cards per year and is one of the busiest in the world. The Society has a policy of discouraging the sending of cards when they are not wanted or cannot be received.

Active Amateurs, GB and Special events, all Clubs and DXpeditions are strongly advised that 100% QSL outgoing is no longer desirable or cost effective.

Transporting large volumes of cards between bureaus, only to have them ultimately destroyed, returned or uncollected, is disappointing and not eco-friendly.

Tip: Ask yourself… Do I need to send a card for every contact before QSLing? Always ask the other station if they can receive a bureau card, before sending.

Log Book of the World (LOTW)

Receiving a nice card for a memorable contact is always a thrill, never matched by an electronic confirmation via the Internet. However, do consider the alternatives, uploading your logs to Logbook of The World can automatically confirm some contacts, such as for contests and award purposes etc.

Confirmations via LOTW are easy and work well for everyone, if a few simple steps are followed. See: www.arrl.org/logbook-of-the-world

OQRS systems - the future of QSLing?

Many stations and most DXpeditions and rare calls are now using the worldwide OQRS network and only responding to requests for QSL cards. This online system means there is now no need to automatically send a card, to receive one via the bureau, or direct. Using OQRS also speeds up the system so that it can now be only half the time it presently takes to send and receive a card, with the added benefit of not needing to send yours. Simply put cards are only sent in response to OQRS requests for a card. So if you are sending QSL cards you or your QSL manager will receive an email to generate a genuinely wanted card. This saves time and waste for both the user and the QSL system in general and is therefore recommended as good practice.

In the UK we are fortunate to have the free to use ClubLog, courtesy of Michael Wells, G7VJR and his team. Simply go to, www.clublog.org for more information or to register your call, club, GB station or event and start uploading your logs

Hand written cards can be easily pre-sorted, using a simple card index box.

Sending cards via the Bureau

Cards from RSGB members for both UK and worldwide should be sent, suitably packed, to the main UK bureau address: **RSGB QSL Bureau PO Box 5, Halifax HX1 9JR, England.**

Members, clubs or DX groups wishing to send large or heavy packages to the Bureau via carriers other than Royal Mail should contact the Bureau for an alternative delivery address.

Responsible QSLer…

Help us to speed up processing and cut waste for everyone,

- 10 Simple things you can do.....
- Ask new contacts, "If I send you a QSL card, do you collect and how? – every time
- If you don't QSL, be polite but honest "Thanks but no thanks" is all it takes.
- Please don't say you do when you don't, or ignore the other guy's kind offer,
- QSLing 100% outgoing is costly for everyone. Half your cards could be wasted, please check before you send.
- Create your own OQRS system or use ClubLog www.clublog.org to reply to incoming requests with a real card, save time and money for you or your club.
- Make your QRZ.com QSL details clear, honest and visible
- Amend your on-line details if you change your QSL status - don't leave it.
- QSL info– Direct or via…' is confusing. Please clarify what you really mean .
- Clubs Calls – Collect only from the calls sub group. For, 'QSL-Direct,' show an address not a call-sign as this can change, it's often confusing .
- Always collect your cards. - Even if you never send one, they will arrive.

Be Responsible, it only costs a stamp!

Fair Usage Policy

As part of their subscription, each Member can send up to 15 Kgs of cards through the Bureau each year (about 5000 standard cards).

Each Affiliated Club can send up to 20 Kgs through the Bureau each year. Additional cards will be charged at £6 per kilo or part thereof.

In the interests of fairness to others, members should only send to the Bureau a maximum of 1kgs of cards (approx 300) to any single DXCC entity per month (larger quantities should be sent directly to the bureau in the relevant country - *see* IARU list online).

Heavy users such as DXpeditions, some clubs, etc, will be required to send the bulk of their outgoing cards direct to the destination countries.

Each batch of cards should contain...

- Proof of current membership; that is an original *RadCom* address label showing address, callsign and membership number, not more than three months old. Leaving out this information will likely result in a delay, while membership status is checked.
  As Clubs receive the RSGB Yearbook each year in lieu of RadCom, they should include sufficient information for a check to be made against the Affiliated society's register. Ideally, in the form of a club letterhead, showing the membership number and renewal date. To speed status checking, clubs and groups are asked to ensure that they register club call and contact details at *My Account* directly in the group's name and not as secondary to a personal call sign, or qsl manager.
- Special event stations (GB) and single letter Abbreviated/Contest callsigns should include the membership number and call of the NoV license holder or affiliated club, for contact purposes.

Other important points

- Clubs and QSL Managers sending a bulk dispatch to the bureau should ensure that all callsign holders for whom they send cards are current members of RSGB and should enclose current membership details for every callsign with every batch of cards.
- Members who operate from another station, typically a foreign club call or that of an individual overseas amateur, may send cards for contacts made from that station, provided they clearly identify themselves as the operator and state their UK callsign and membership number on each card.
- Listener report QSLs need sufficient information to be of genuine value to the transmitting amateurs. Reception reports relating to broadcasting stations cannot be accepted.
  The bureau system accepts standard cards only no letters, SAE's or money orders.
- All cards, whatever the quantity, should be pre-sorted into alphabetical and numerical country DXCC order (see the Prefix List pages and/or the *RSGB Prefix Guide* which also contains a complete cross

reference and awards section).

Countries with more than one prefix should be placed together. For example, JA and 7J cards (destined for Japan), F and TK-TM cards (destined for France) and SP, HF and 3Z (destined for Poland) may be grouped together.

Cards for the USA need be sorted separately into call areas (numbers 0-9), regardless of the prefix letters.

NB: Exceptionally, cards in the number 4 series with either one or two letters before the number are handled by different bureaus and need to be separated, as are cards for Alaska (KL), Hawaii (KH6-7) and Puerto Rico (KP3-4).

Cards for Russian Federation and former Soviet countries were traditionally grouped together. They now need to be separated into five individual groups; RA-RZ, UA-UI, UJ-UM, UN-UQ and UR-UZ, as they are no longer sent to a single destination in Moscow. See: *www.iaru.org/iaruqsl.html*

Cards for UK delivery need to be provided separately to foreign destinations. Our UK sorters currently have to split cards into some 112 alphanumeric categories. For this reason cards should be supplied to us pre-sorted order as per the Sub Manager list on the following pages.

Envelopes, paper or card dividers to separate countries or call groups are not required, as removing these can sometimes slow down the distribution process. Cards sent in date/time/logbook or random order are not acceptable, as they typically take up to five times longer to process. Similarly, those with small print or hand written callsigns can be very difficult to process, resulting in delays for other users. The bureau reserves the

right, at its discretion, to reject unsorted cards or those with callsign or routing information in small and difficult to read print. The minimum print size requirement is 12 point.

Tip: If you are unsure about your handwriting, why not ask someone else to check the cards to see if they can easily read the callsigns?

Card issues and some good advice

To avoid possible transit damage and in fairness to others, all cards should be single page, standard postcard sized (140 x 90mm). Card weight/thickness is important and needs

---

Checklist for sending cards

We need your help to sort more than a million cards each year and reduce delays.

A   *First, place your cards into three piles...*

1. UK destinations
Pre-sort G, M and 2 as per the Sub Managers list.

2. USA destinations
Sort by number only, 0-9, regardless of prefix. Separate cards for Alaska, Hawaii and Puerto Rica.

3. Rest of the World
In DXCC callsign prefix order.

4. Calls with numbers first
Sort in digit order, i.e. 3A, 4X, 8P, 9H etc.

5. Calls with one letter then one number

These come before two letter prefixes, i.e. S5 before SM, etc.

B   *Check ALL cards for possible 'Via' destinations*

Re-sort if necessary, and never rely on your computer print log, for example: F5/G3UGF isn't a French destination.

Africa, Caribbean and DX destinations are mostly QSL Direct only, or via a QSL Manager. Check *www.qrz.com*

C   *Pack your cards securely and don't forget*

A recent RadCom address wrapper as proof of membership.
Your callsign and return address on the package.
If you put more than ten cards in a C5 envelope, check the dimensions and weight at the Post Office, before sending don't just post.

Whatever the quantity, never send unsorted cards!

Before you send, please check that we can deliver!

Check all Vias before you send, as your card may come back to you, or it may never arrive.

There are many world destinations, but only 190 IARU member and associate Bureaus worldwide.

The following IARU Bureaus are currently closed.

3B Mauritius	D4 Cape Verde
3DA Swaziland	HH 4V Haiti
4J-K Azerbaijan	HV Vatican City
7P Lesthoto	PZ Suriname
9L Sierra Leone	ST Sudan
A3 Tonga	SU, 6A-B Egypt
C2 Nauru	V3 Belize
C5 Nauru	V4 St Kitts and Nevis
C6 Bahamas	XY-XZ Myanmar
CN, 5C-G Morocco	Z2 Zimbabwe

Download your own Bureau list from:
www.iaru.org/iaruqsl.html

to be in the range 130-330gm paper board for easy processing.

Large or unusual shaped cards are extremely difficult to process and most easily damaged when packed, or folded with others.

Thin, small, and paper cards are slow and extremely difficult to handle. They often stick to cards for other destinations, as do homemade cards using photo print or heat laminated paper. This type of card does not travel well, is difficult to write on and is very easily damaged when subjected to humidity or damp – not recommended!

Multiple page and non-standard cards should be avoided, as they increase the Society's workload and overheads, at the expense of other members. They can also significantly reduce the numbers of individual cards per consignment to overseas destinations. In quantity they should be sent direct to overseas bureaus. Therefore, in the interest of fairness to other members, they are sometimes spread over several shipments.

If multiple page cards are used they should weigh no more than a single page standard card typically 3gm maximum and should be pre folded to clearly show the destination callsign.

QSL Routing & QSL Direct

It sounds obvious, but the Bureau can only process outgoing cards if there is a destination to which they can be sent. Before sending cards, therefore, (particularly to rare stations or DXpeditions) please check the recipient's QSL policy. This is usually available on QRZ. com or via a websearch.

Many DXpeditions and rare call signs only QSL direct, or respond to an OQRS request often via a QSL Manager who may be in another country. These stations are most often located where there is no bureau service and are operated by visiting non-resident Amateurs from another country. Some stations do not QSL at all, so it is vital to check before sending, whatever route you choose.

Please note that outgoing bureau cards where no destination bureau is available, or no clear 'Via' route is indicated, will be recycled.

Tip:

- Ask for the other station's QSL details at the time of the QSO, or by an Internet search before posting.

- Consider posting your most wanted cards direct or to an overseas bureau, if it's active. This helps to speed replies as most bureaux world-wide have backlogs. The IARU world-bureau list can be downloaded at: *www.iaru.org/ qsl-bureas.html*. It's good practice to check the listings for changes at least twice a year.

- Always search the web and check *www. qrz.com* first before posting.

- Make sure that any "via" information on your cards appears directly below, or next to, the station call sign, to avoid being missed. Using a different coloured ink for this purpose is a great help.

For guidance on what information to include

on your card (and where), *see the example card on page 21.*

Using printed labels

Avoid cramming too much information on small printed labels. For health and safety reasons all callsigns should be a minimum of 12 point print size and in common, easy to read fonts such as Arial, Times New Roman, or clear, block capital hand written letters.

The bureau system is for the exchange of QSL cards only. Envelopes containing letters, photographs, IRCs, stamped addressed envelopes, awards, certificates and other items will not be processed and should be sent by other means.

Heavy users

Those sending more than a few thousand cards per year should send their largest volumes of cards directly to their top ten destination countries. The remaining balance can be sent via the normal bureau system.

The aim of this is to share some of the burden of cost, without penalising others who may only occasionally send a few more cards than normal. The bureau weighs and notes regular large consignments. Members or clubs may be contacted if their usage becomes excessive, with a request to follow the guidelines above. The IARU bureau list can be found at: *www.iaru.org/iaruqsl.html*

Packing and posting your cards

The bureau receives many damaged envelopes and packages from both UK and foreign amateurs. It also receives a significant number of requests each month from Royal Mail for payment of additional postage, which are always rejected.

Having first separated pre-sorted, UK destination cards from the rest of the world, please read on...

- Never post loose cards in lightweight or thin envelopes, as they will often cut through the edge of the envelope in transit.

- Always print return address details and callsign on your package, in case it arrives damaged.

- Secure batches of cards with a rubber band or - better still - a banknote style band of thin paper strip, folded around the cards.

- Never place two or more packs of cards side by side in a C5, A4 or larger envelope, as it will fold in transit and split down the middle, allowing the cards to spill out.

- Using lightweight 'Mail-Tuff' style plastic or 'Mail-Lite' style padded bags or Post-Pack envelopes usually avoids this problem.

- Always check the size and weight of your envelopes and packages, before posting as the Post Office now charge by volume as well as by weight.

- The current weight limit for a First Class stamp is 100g, but the package size is limited to 240mm x 165mm and the package should fit through a postal slot only 5mm in height.

- It is possible to send a large envelope A4, or a smaller envelope over 5mm in thickness). This type of envelope is considered to be a 'Large Letter.' Large Letter stamps costs more, but allows the letter thickness to

Part of the overseas side of the bureau

be up to 25mm. 'Second Class Large' offers better value

- The Post Office can supply a paper/card copy of their pricing slot guide for a small charge. Frequent users are advised to obtain a plastic Helix HP5 'Pricing in Proportion' Ruler. It has postal slots built in, to check your packets.

Sending small numbers of cards in separate envelopes is not cost effective for the sender and means much more time spent opening, sorting and checking in the bureau. Sending not more than one pack per month, with your *RadCom* label, resolves many issues and can save you money .

Recorded delivery is not cost effective. We receive many packages and we are not always asked to sign for individual items, secure packing and a return address offers better value.

Receiving cards from the Bureau

RSGB is extremely fortunate to have around 80 dedicated volunteer Sub Managers who give freely of their time to support the work of the Bureau and in the service of their fellow radio amateurs. Members' cards are sent to the Sub Managers for onward distribution.

Sub Managers details are subject to change, so it a good idea to check the QSL section of the RSGB website from time to time for the latest information. From the RSGB Home Page click, 'Operating' and follow the links.

Our system relies upon those wishing to receive cards depositing Stamped Addressed Envelopes with Sub Managers, ready for each quarterly despatch.

Members should use SAEs, as Sub Managers are not authorised or insured to accept money in lieu of postage stamps. RSGB is not liable in case of any loss or dispute.

The scheme is open to all RSGB members plus UK-based, pay-to-receive subscribers.

Collection Envelopes

- Envelopes need to be C5 size (160mm x 230mm) and of strong material (*see earlier*).

- Callsign or Listener number should be printed in the top left hand corner, followed by the a current membership number, immediately below.

- Print the name, delivery address and postcode clearly, as normal.

- Number each envelope sent to the Manager (eg '1 of 6', '2 of 6', '3 of 6', etc)

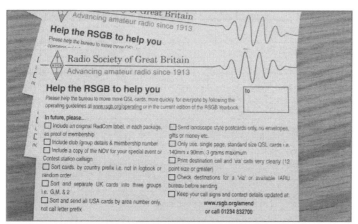

always mark one of them 'Last', so that you will know when a fresh batch should be sent.

- Envelopes are normally despatched every quarter, subject to card availability.
- Always use stamps worded Second or First Class, rather than a numerical amount, as these will be honoured if the postal rate changes.

No delivery in that quarter means 'no cards waiting'.

Cards for amateurs who have not lodged envelopes are not returned to sender and will, at the Sub Manager's discretion, be recycled after a period of three months. Always keep your envelopes up to date. Many volunteer Sub Managers now operate their own websites, with links from the RSGB website, giving cards waiting, envelope status and next anticipated delivery details. RSGB requires these lists to be confidential. Members permission to display their callsign and details to others, is a condition of inclusion on any such listing, operated by a volunteer.

It is a good idea to note in your diary to check your Manager's list every quarter.

UK amateurs who do not wish to collect cards or those who use a separate QSL Manager are asked to notify the appropriate Sub Manager as a matter of courtesy and also make this clear at www.qrz.com

More than one callsign?

Stations changing their callsign as a result of a licence upgrade etc should inform RSGB of their change of status, using the 'Amend my Details' form accessed from the home page of the RSGB website, (see 'Operating' listings) or by telephoning headquarters. They also need to maintain envelopes with both the new and old QSL Sub Managers. Typically, envelopes for the old callsigns and membership number need to be available for up to five years after the old call is no longer the primary call.

Club stations should enter their callsign details in the club's name and not as a secondary call of a member or QSL manager, as this gives rise to confusion. Please avoid registering calls using optional club identifiers, such as X,S,C,N

Stations operating from a different prefix, for example G9ABC as GW9ABC/P or GU9ABC/P, need to lodge envelopes with the appropriate Sub Managers for every area of operation, as cards may not be forwarded to the home call.

UK mainland stamps are not valid when sent from the Isle of Man or Channel Islands. Local stamps should be obtained during the period of operation, for use later. When operating outside the UK under CEPT rules, e.g. F/G9ABC, or more importantly with another callsign, it is vital to tell the QSO partner to 'QSL via G9ABC' and not simply state 'via home call'.

Registering any foreign calls separately, together with the QSL route and contact email address at: *www.qrz.com* is extremely helpful to others in these cases.

Requesting a 'Via' call route

In recent years there has been an explosion in the use of 'Via' requests, where amateurs use a QSL Manager, or wish to have cards sent c/o another callsign. Advising your contacts to send cards via the personal callsign of the RSGB volunteer Sub Manager is not appropriate, as he/she may change.

In many instances the incoming card does not contain the 'Via' information given during the QSO, the expectation being that the bureau sorters will instinctively know the routing.

With so many cards passing through the bureau each week - and the passage of time - this is simply not practical or possible. Finding the routing is a time-consuming process and no longer a realistic or reliable option for our staff.

This problem can be easily avoided by lodging SAEs with the correct Sub Manager, for the actual call used but bearing any alternative delivery address, i.e. that of the station's QSL Manager. This should be considered as a more effective solution and is much to be preferred over giving out a QSL Manager or 'Via' details to every contact.

For example, 'G9ABC, QSL via M8ZZZ' can simply be replaced by sending all cards to the G9ABC RSGB Sub Manager, who holds envelopes marked with the street address for M8ZZZ.

In the case of special event (GB calls), together with all Abbreviated/Contest (single letter suffix) callsigns and personal Special Prefix calls (GR. MQ, 2O. MV etc), no Vias are accepted.

All bureau cards for these groups are sent directly to specialist Sub Managers - see list. These Managers will only send cards to the NoV holder, unless an authorised alternative destination is confirmed in writing by the callsign holder and registered via the website.

Remember: Even if you never send a QSL card, someone somewhere, sometime, will send you a card. It would be a shame not to receive it, so please send an SAE to your RSGB volunteer sub manager!

Card design

Whether you are designing and making your own QSL or having it made professionally, *size, quality and design are the most important factors* if you are hoping for a reply.

Gone are the days when cards were printed in a single colour (black), with only a callsign and basic information and which took several weeks to produce. The advent of high resolution digital photography and computers has changed everything. High quality commercial QSL cards are now more interesting, colourful, easier, quicker and much cheaper to produce or change than ever before. What's more, professionally printed cards can and often do work out cheaper than making your own.

Checklist for receiving cards

1. Register all your callsigns, past/present

Do this via the RSGB website at 'amend my details', or telephone 01234 832700.

2. Send C5 stamped envelopes to each Sub Manager

In addition to your name and address...
Write your callsign and RSGB membership number at the top left.
Number each envelope at the bottom left.
Mark 'Last envelope' at the bottom left of the last envelope.
N.B. Sub Managers are not authorised to accept cash in lieu of SAEs.

3. Holidays and portable activations

We don't automatically divert cards to your home call.
For temporary prefix operation (e.g. GM, MW, 2I etc), lodge separate envelopes with the relevant Sub Manager to collect your cards.
N.B. The Channel Islands and Isle of Man use different stamps.

4. Special event (GB) and abbreviated contest calls (G1A, etc)

No diversions apply, so please see the Sub Manager list.

5. Special Prefix NoV callsigns

For GR, MQ, 2O etc, no diversions apply. See the Sub Manager details.
Multiple callsigns can be listed on the same envelope.

G4EZT

CONFIRMING OUR QSO(s):

TO: 4K9W
VIA: DL 6 KVA

DATE			TIME	FREQ/MHz	MODE/2x	RST
DAY	MONTH	YEAR	GMT			
29	12	2007	12:31	14·079	RTTY	599
16	10	2007	19:18	10·138	RTTY	599
18	11	2006	10:43	18·145	SSB	59
3	10	2004	10:03	24·945	SSB	57

TCVR: IC756 + AL-572
ANT: Yagi / Vertical / Doublet

PSE / QSL VIA / RSGB DIRECT
TNX

ANDY P. BROWN
"THE FIRS", WINDMILL HILL, TEMPLE GRAFTON, STRATFORD-UPON-AVON, WARKS, B50 4LH, ENGLAND.

73 de Andy

Example of a QSL card that's well laid out, easy to read and easy to sort.

All the more reason to consider having a distinctive card that gives not just your station details, but perhaps reflects your radio and other interests, family, pets, location or some other part of your life. Cards can be simple, beautiful, artistic, funny, technical or even something completely unexpected. They make a statement about you - so what does your card say?

The range of choice has never been greater, so just use your imagination. Above all, make your QSL card something of quality that stands out; something that the other station will want to keep and display. If you are sending or receiving a 'gift', make it memorable. It's now possible to collect special interest cards showing planes, trains, ships, cars, families, pets, castles, churches, windmills, lighthouses, motorcycles and many other things, in addition to antenna farms, radios, vintage gear and shack interiors.

Tip: Remember to tidy up before you take a photo of your station!

The business side of the card is also very important. Here, simple clarity is the key to a good card and to receiving a reply. Use a clear type face that is easy to read. Don't put too much information or too many logos on the card, unless it's a special event when background information is always nice to see. Remember that English is not always a first language, so keep it simple, keep it relevant. Allow enough space to write or print the contact information clearly on the card, ensuring that the destination call is at the **top right, with any via routing details immediately below.**

Many cards now have space to log more than one contact. This is a great eco-friendly idea. *See example card below from G4EZT.*

Where to buy cards

The RSGB doesn't endorse any particular producer. Take a look at the cards you receive, as they will often include maker's details.

Apart from your local printer and checking with friends, there are now a whole range of specialist on-line makers offering superb, correctly sized card. We regularly see cards from UK stations being sent to us that have been designed on-line, some produced in other countries, and many are simply stunning. It is possible to download card making software from the Internet, but so much depends on the actual equipment used to make the card that the results are often disappointing or uneconomic, unless you have access to specialist print and cut-ting machinery. However, where practical, they do make possible one-off special, individual and personalised cards, for QSLing direct.

Remember: If you have invested time and energy on your station, isn't it right to do the same with your QSL card? Send something you would be pleased to receive.

My QSL Sub Manager is:

Sub Manager details can be found on the rsgb website
www.rsgb.org/qsl

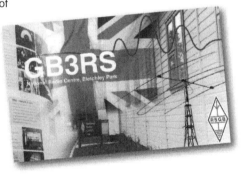

Intruder Watch

The RSGB Monitoring System, more popularly known as the Intruder Watch is a small team of volunteer observers and forms part of the IARU Monitoring System. As such it submits reports of non-amateur transmissions heard on the exclusive HF amateur bands to both the Ofcom Monitoring Station at Baldock and the IARU Region 1. While most of Intruder Watch activities is centred around the HF bands, Intruder Watch also assists leasing with Ofcom and AROS for reports of non-amateur transmissions in the VHF/UHF bands.

Intruders removed from our exclusive amateur bands include broadcast stations, military data transmissions, faulty positioning installations, coast stations, embassies, fax stations, faulty set-top boxes and numerous others. For data transmissions a 'zero beat' frequency will be accurate enough for our observers without decoders. Many software data decoders are available as well as equipment manufactured by companies such as 'Hoka', 'Wavecom' and 'Universal' these are used for analysis of data signals.

Most information received by the co-ordinator arrives from regular observers, but occasional reports are also welcome from anyone who finds what may be an intruding signal on one of our exclusive amateur bands. This information can then be passed on to a suitably equipped observer for further investigation. All reports are welcome and will be acknowledged.

Data communications is by far the most common intruder into the HF amateur bands and it is an area where we could use more support. Other non-data categories of intruding signals include CW, broadcast stations, speech and over-the-horizon radar (OTHR). Any report should include as much information as possible, but preferably frequency, date, time (UTC), mode of transmission, any identification signal or callsign, language used, text (where appropriate) and beam heading where possible.

Intruder Watch is always looking for more volunteer observers so if you think you might like to join our team do please send an email to the Intruder Watch Co-ordinator Vaughan Ravenscroft, M0VRR.
Email: iw@rsgb.org.uk

Contest Calendars

Contests are sporting events between amateur stations on specific bands and modes, conducted according to published rules. The activity appeals mainly to those with a competitive instinct, but construction and station optimisation are also important.

2018 RSGB HF Contest Calendar

Date (2018)	Time	Contest Name	Sections
Sun 14 Jan	1400-1800	RSGB AFS 80m-40m Contests CW	400W 100W 10W
Sat 20 Jan	1400-1800	RSGB AFS 80m-40m Contests Phone	400W 100W 10W
Mon 5 Feb	2000-2130	80m CC SSB	100W 10W
Sat 10 Feb	1900-2300	**1st 1.8MHz Contest**	UK-Assisted UK-Unassisted Non-UK-Assisted Non-UK-Unassisted
Wed 14 Feb	2000-2130	80m CC DATA	100W 10W
Thu 22 Feb	2000-2130	80m CC CW	100W 10W
Mon 5 Mar	2000-2130	80m CC DATA	100W 10W
Sat 10-Sun 11 Mar	1000-1000	**Commonwealth Contest**	OPEN-SOA OPEN-SOU RESTRICTED-SOA RESTRICTED-SOUMulti-Op HQ QRP-BERU
Wed 14 Mar	2000-2130	80m CC CW	100W 10W
Thu 22 Mar	2000-2130	80m CC SSB	100W 10W
Sun 1 Apr	1900-2030	**RoLo**	ALL
Mon 2 Apr	1900-2030	80m CC CW	100W 10W
Wed 11 Apr	1900-2030	80m CC SSB	100W 10W
Thu 26 Apr	1900-2030	80m CC DATA	100W 10W
Mon 7 May	1900-2030	80m CC SSB	100W 10W
Wed 16 May	1900-2030	80m CC DATA	100W 10W
Thu 31 May	1900-2030	80m CC CW	100W 10W
Sat 2-Sun 3 Jun	1500-1500	NFD	QRP-Portable Low-Power-Unassisted-Portable Low-Power-Assisted-Portable Fixed
Mon 4 Jun	1900-2030	80m CC DATA	100W 10W
Wed 13 Jun	1900-2030	80m CC CW	100W 10W
Thu 21 Jun	1900-2030	80m CC SSB	100W 10W
Mon 2 Jul	1900-2030	80m CC CW	100W 10W
Wed 11 Jul	1900-2030	80m CC SSB	100W 10W
Sun 22 Jul	0900-1600	Low Power Contest	A B C D
Thu 26 Jul	1900-2030	80m CC DATA	100W 10W
Sat 28-Sun 29 Jul	1200-1200	IOTA Contest	Single Operator Unassisted Single Operator Assisted Multi-Single Multi-Two
Sat 1-Sun 2 Sep	1300-1300	SSB Field Day	QRP-Portable Low-Power-Unassisted-Portable Low-Power-Assisted-Portable Fixed
Mon 10 Sep	1900-2030	Autumn Series SSB	100W 10W
Wed 19 Sep	1900-2030	Autumn Series CW	100W 10W
Thu 27 Sep	1900-2030	Autumn Series DATA	100W 10W
Sun 7 Oct	0500-2300	DX Contest	400W 100W 10W
Mon 8 Oct	1900-2030	Autumn Series CW	100W 10W
Wed 17 Oct	1900-2030	Autumn Series DATA	100W 10W
Sun 21 Oct	1900-2030	RoLo	ALL
Thu 25 Oct	1900-2030	Autumn Series SSB	100W 10W
Sat 10 Nov	2000-2300	Club Calls (1.8MHz AFS)	ALL
Mon 12 Nov	2000-2130	Autumn Series DATA	100W 10W
Sat 17 Nov	1900-2300	**2nd 1.8MHz Contest**	UK-Assisted UK-Unassisted Non-UK-Assisted Non-UK-Unassisted
Wed 21 Nov	2000-2130	Autumn Series SSB	100W 10W
Thu 29 Nov	2000-2130	Autumn Series CW	100W 10W

For Key to HF Special Rules etc. please see overleaf

Events in bold qualify for the HF Championship

The Contest calendars in these pages are provisional, and before entering you should check the website before getting on the air to participate at: *www.rsgbcc.org*

2018 RSGB VHF Contest Calendar

Date	Time (UTC)	Contest Name	Sections
1st Tue of Month	1900-2000 L	144MHz FMAC	FR FL
1st Tue of Month	2000-2230 L	144MHz UKAC	AO AR AL
2nd Tue of Month	1900-2000 L	432MHz FMAC	FR FL
2nd Thur of Month	2000-2230 L	50MHz UKAC	AO AR AL
2nd Tue of Month	2000-2230 L	432MHz UKAC	AO AR AL
3rd Tue of Month	2000-2230 L	1.3GHz UKAC	AO AR AL
3rd Thur of Month	1900-2000 L	70MHz FMAC	FR FL
3rd Thur of Month	2000-2230 L	70MHz UKAC	AO AR AL
4th (Jan-Nov)	2000-2230 L*	SHF UKAC SAO SAR	
4 Feb	0900-1300	432MHz AFS Super League AFS Super League	O SF AFS
25 Feb	1000-1200	70MHz Cumulatives #1	O SF
3- 4 Mar	1400-1400	**March 144 432MHz VHF Championship**	O 6O SF SO 6S
11 Mar	1000-1200	70MHz Cumulatives #2	O SF
1 Apr	0900-1200	First 70MHz Contest	O SF
15 Apr	0900-1200	First 50MHz Contest	O SF
5 May	1400-2200	10GHz Trophy Contest	A
5 May	1400-2200	**432MHz Trophy Contest VHF Championship**	O SF
5- 6 May	1400-1400	May 432MHz-245GHz Contest	O SF
13 May	0900-1200	70MHz Contest CW VHF CW Championship	O SF
19- 20 May	1400-1400	**144MHz May Contest VHF Championship**	O SF SO 6S 6O
20 May	1100-1500	1st 144MHz Backpackers	5B 25H
27 May	1400-1600	70MHz Cumulatives #3	O SF
10 Jun	0900-1300	2nd 144MHz Backpackers	5B 25H
16- 17 Jun	1400-1400	**50MHz Trophy Contest VHF Championship**	O SF SO 6O 6S Overseas
24 Jun	1400-1600	70MHz Cumulatives #4	O SF
24 Jun	0900-1200	50MHz Contest CW VHF CW Championship	O SF
7- 8 Jul	1400-1400*	VHF NFD	Open R L MMS FSO FSR
8 Jul	1100-1500	3rd 144MHz Backpackers	5B 25H
22 Jul	1000-1600	**70MHz Trophy Contest VHF Championship**	O SO SF
4 Aug	1300-1700	4th 144MHz Backpackers	5B 25H
4 Aug	1400-2000	144MHz Low Power Contest VHF Championship	O SF SO
5 Aug	0800-1200	**432MHz Low Power Contest VHF Championship**	O SF SO
12 Aug	1400-1600	70MHz Cumulatives #5	O SF
1- 2 Sep	1400-1400	**144MHz Trophy Contest VHF Championship**	O SF SO 6O 6S
2 Sep	1100-1500	5th 144MHz Backpackers	5B 25H
16 Sep	0900-1200	Second 70MHz Contest	O SF
6- 7 Oct	1400-1400	Oct 432MHz-245GHz Contest	O SF
6 Oct	1400-2200	**1.2GHz Trophy / 2.3GHz Trophy VHF Championship**	O SF
21 Oct	0900-1300	50MHz AFS Contest AFS Super League AFS Super League	O SF
3- 4 Nov	1400-1400	144MHz CW Marconi VHF CW Championship	SF O 6S 6O
2 Dec	1000-1600	144MHz AFS AFS Super League AFS Super League	SF O
26-29 Dec	1400-1600	50/70/144/432MHz Christmas Cumulatives Contest	SF O

Events in bold qualify for the VHF Championship

*Times marked with a * denote times vary per band   L = Local*

NOTES

2018

DIARY

2018

January
M	T	W	T	F	S	S
1	2	3	4	5	6	7
8	9	10	11	12	13	14
15	16	17	18	19	20	21
22	23	24	25	26	27	28
29	30	31				

February
M	T	W	T	F	S	S
			1	2	3	4
5	6	7	8	9	10	11
12	13	14	15	16	17	18
19	20	21	22	23	24	25
26	27	28				

March
M	T	W	T	F	S	S
			1	2	3	4
5	6	7	8	9	10	11
12	13	14	15	16	17	18
19	20	21	22	23	24	25
26	27	28	29	30	31	

April
M	T	W	T	F	S	S
						1
2	3	4	5	6	7	8
9	10	11	12	13	14	15
16	17	18	19	20	21	22
23	24	25	26	27	28	29
30						

May
M	T	W	T	F	S	S
	1	2	3	4	5	6
7	8	9	10	11	12	13
14	15	16	17	18	19	20
21	22	23	24	25	26	27
28	29	30	31			

June
M	T	W	T	F	S	S
				1	2	3
4	5	6	7	8	9	10
11	12	13	14	15	16	17
18	19	20	21	22	23	24
25	26	27	28	29	30	

July
M	T	W	T	F	S	S
						1
2	3	4	5	6	7	8
9	10	11	12	13	14	15
16	17	18	19	20	21	22
23	24	25	26	27	28	29
30	31					

August
M	T	W	T	F	S	S
		1	2	3	4	5
6	7	8	9	10	11	12
13	14	15	16	17	18	19
20	21	22	23	24	25	26
27	28	29	30	31		

September
M	T	W	T	F	S	S
					1	2
3	4	5	6	7	8	9
10	11	12	13	14	15	16
17	18	19	20	21	22	23
24	25	26	27	28	29	30

October
M	T	W	T	F	S	S
1	2	3	4	5	6	7
8	9	10	11	12	13	14
15	16	17	18	19	20	21
22	23	24	25	26	27	28
29	30	31				

November
M	T	W	T	F	S	S
			1	2	3	4
5	6	7	8	9	10	11
12	13	14	15	16	17	18
19	20	21	22	23	24	25
26	27	28	29	30		

December
M	T	W	T	F	S	S
					1	2
3	4	5	6	7	8	9
10	11	12	13	14	15	16
17	18	19	20	21	22	23
24	25	26	27	28	29	30
31						

January 2018

1	Monday	
2	Tuesday	
3	Wednesday	
4	Thursday	
5	Friday	
6	Saturday	
7	Sunday	
8	Monday	
9	Tuesday	
10	Wednesday	
11	Thursday	
12	Friday	
13	Saturday	
14	Sunday	
15	Monday	
16	Tuesday	
17	Wednesday	
18	Thursday	
19	Friday	
20	Saturday	
21	Sunday	
22	Monday	
23	Tuesday	
24	Wednesday	
25	Thursday	
26	Friday	
27	Saturday	
28	Sunday	
29	Monday	
30	Tuesday	
31	Wednesday	

February 2018

Thursday	1
Friday	2
Saturday	3
Sunday	4
Monday	5
Tuesday	6
Wednesday	7
Thursday	8
Friday	9
Saturday	10
Sunday	11
Monday	12
Tuesday	13
Wednesday	14
Thursday	15
Friday	16
Saturday	17
Sunday	18
Monday	19
Tuesday	20
Wednesday	21
Thursday	22
Friday	23
Saturday	24
Sunday	25
Monday	26
Tuesday	27
Wednesday	28

March 2018

1	Thursday	
2	Friday	
3	Saturday	
4	Sunday	
5	Monday	
6	Tuesday	
7	Wednesday	
8	Thursday	
9	Friday	
10	Saturday	
11	Sunday	
12	Monday	
13	Tuesday	
14	Wednesday	
15	Thursday	
16	Friday	
17	Saturday	
18	Sunday	
19	Monday	
20	Tuesday	
21	Wednesday	
22	Thursday	
23	Friday	
24	Saturday	
25	Sunday	
26	Monday	
27	Tuesday	
28	Wednesday	
29	Thursday	
30	Friday	
31	Saturday	

April 2018

	Sunday	1
	Monday	2
	Tuesday	3
	Wednesday	4
	Thursday	5
	Friday	6
	Saturday	7
	Sunday	8
	Monday	9
	Tuesday	10
	Wednesday	11
	Thursday	12
	Friday	13
	Saturday	14
	Sunday	15
	Monday	16
	Tuesday	17
	Wednesday	18
	Thursday	19
	Friday	20
	Saturday	21
	Sunday	22
	Monday	23
	Tuesday	24
	Wednesday	25
	Thursday	26
	Friday	27
	Saturday	28
	Sunday	29
	Monday	30

May 2018

1	Tuesday	
2	Wednesday	
3	Thursday	
4	Friday	
5	Saturday	
6	Sunday	
7	Monday	
8	Tuesday	
9	Wednesday	
10	Thursday	
11	Friday	
12	Saturday	
13	Sunday	
14	Monday	
15	Tuesday	
16	Wednesday	
17	Thursday	
18	Friday	
19	Saturday	
20	Sunday	
21	Monday	
22	Tuesday	
23	Wednesday	
24	Thursday	
25	Friday	
26	Saturday	
27	Sunday	
28	Monday	
29	Tuesday	
30	Wednesday	
31	Thursday	

June 2018

	Friday	1
	Saturday	2
	Sunday	3
	Monday	4
	Tuesday	5
	Wednesday	6
	Thursday	7
	Friday	8
	Saturday	9
	Sunday	10
	Monday	11
	Tuesday	12
	Wednesday	13
	Thursday	14
	Friday	15
	Saturday	16
	Sunday	17
	Monday	18
	Tuesday	19
	Wednesday	20
	Thursday	21
	Friday	22
	Saturday	23
	Sunday	24
	Monday	25
	Tuesday	26
	Wednesday	27
	Thursday	28
	Friday	29
	Saturday	30

July 2018

1	Sunday	
2	Monday	
3	Tuesday	
4	Wednesday	
5	Thursday	
6	Friday	
7	Saturday	
8	Sunday	
9	Monday	
10	Tuesday	
11	Wednesday	
12	Thursday	
13	Friday	
14	Saturday	
15	Sunday	
16	Monday	
17	Tuesday	
18	Wednesday	
19	Thursday	
20	Friday	
21	Saturday	
22	Sunday	
23	Monday	
24	Tuesday	
25	Wednesday	
26	Thursday	
27	Friday	
28	Saturday	
29	Sunday	
30	Monday	
31	Tuesday	

August 2018

	Wednesday	1
	Thursday	2
	Friday	3
	Saturday	4
	Sunday	5
	Monday	6
	Tuesday	7
	Wednesday	8
	Thursday	9
	Friday	10
	Saturday	11
	Sunday	12
	Monday	13
	Tuesday	14
	Wednesday	15
	Thursday	16
	Friday	17
	Saturday	18
	Sunday	19
	Monday	20
	Tuesday	21
	Wednesday	22
	Thursday	23
	Friday	24
	Saturday	25
	Sunday	26
	Monday	27
	Tuesday	28
	Wednesday	29
	Thursday	30
	Friday	31

September 2018

1	Saturday	
2	Sunday	
3	Monday	
4	Tuesday	
5	Wednesday	
6	Thursday	
7	Friday	
8	Saturday	
9	Sunday	
10	Monday	
11	Tuesday	
12	Wednesday	
13	Thursday	
14	Friday	
15	Saturday	
16	Sunday	
17	Monday	
18	Tuesday	
19	Wednesday	
20	Thursday	
21	Friday	
22	Saturday	
23	Sunday	
24	Monday	
25	Tuesday	
26	Wednesday	
27	Thursday	
28	Friday	
29	Saturday	
30	Sunday	

October 2018

Monday	1
Tuesday	2
Wednesday	3
Thursday	4
Friday	5
Saturday	6
Sunday	7
Monday	8
Tuesday	9
Wednesday	10
Thursday	11
Friday	12
Saturday	13
Sunday	14
Monday	15
Tuesday	16
Wednesday	17
Thursday	18
Friday	19
Saturday	20
Sunday	21
Monday	22
Tuesday	23
Wednesday	24
Thursday	25
Friday	26
Saturday	27
Sunday	28
Monday	29
Tuesday	30
Wednesday	31

November 2018

1	Thursday	
2	Friday	
3	Saturday	
4	Sunday	
5	Monday	
6	Tuesday	
7	Wednesday	
8	Thursday	
9	Friday	
10	Saturday	
11	Sunday	
12	Monday	
13	Tuesday	
14	Wednesday	
15	Thursday	
16	Friday	
17	Saturday	
18	Sunday	
19	Monday	
20	Tuesday	
21	Wednesday	
22	Thursday	
23	Friday	
24	Saturday	
25	Sunday	
26	Monday	
27	Tuesday	
28	Wednesday	
29	Thursday	
30	Friday	

December 2018

	Saturday	1
	Sunday	2
	Monday	3
	Tuesday	4
	Wednesday	5
	Thursday	6
	Friday	7
	Saturday	8
	Sunday	9
	Monday	10
	Tuesday	11
	Wednesday	12
	Thursday	13
	Friday	14
	Saturday	15
	Sunday	16
	Monday	17
	Tuesday	18
	Wednesday	19
	Thursday	20
	Friday	21
	Saturday	22
	Sunday	23
	Monday	24
	Tuesday	25
	Wednesday	26
	Thursday	27
	Friday	28
	Saturday	29
	Sunday	30
	Monday	31

NOTES

LOG
SECTION

Amateur Radio

Date	Time (UTC)		Frequency	Mode	Power	Station
	Start	Finish	(MHz)		(dBW)	called/worked

RSGB - Working for the Future of Amateur Radio

Station Log

Report		QSL		Remarks
sent	received	sent	rcvd	

Amateur Radio

Date	Time (UTC)		Frequency (MHz)	Mode	Power (dBW)	Station called/worked
	Start	Finish				

Station Log

Report		QSL		Remarks
sent	received	sent	rcvd	

Amateur Radio

Date	Time (UTC)		Frequency	Mode	Power	Station
	Start	Finish	(MHz)		(dBW)	called/worked

Station Log

Report		QSL		Remarks
sent	received	sent	rcvd	

Amateur Radio

Date	Time (UTC)		Frequency (MHz)	Mode	Power (dBW)	Station called/worked
	Start	Finish				

Station Log

Report		QSL		Remarks
sent	received	sent	rcvd	

Amateur Radio

Date	Time (UTC)		Frequency (MHz)	Mode	Power (dBW)	Station called/worked
	Start	Finish				

RSGB - Working for the Future of Amateur Radio

Station Log

Report		QSL		Remarks
sent	received	sent	rcvd	

Amateur Radio

Date	Time (UTC)		Frequency	Mode	Power	Station
	Start	Finish	(MHz)		(dBW)	called/worked

RSGB - Working for the Future of Amateur Radio

Station Log

Report		QSL		Remarks
sent	received	sent	rcvd	

Amateur Radio

Date	Time (UTC)		Frequency (MHz)	Mode	Power (dBW)	Station called/worked
	Start	Finish				

RSGB - Working for the Future of Amateur Radio

Station Log

Report		QSL		Remarks
sent	received	sent	rcvd	

Amateur Radio

Date	Time (UTC)		Frequency (MHz)	Mode	Power (dBW)	Station called/worked
	Start	Finish				

Station Log

Report		QSL		Remarks
sent	received	sent	rcvd	

Amateur Radio

Date	Time (UTC)		Frequency	Mode	Power	Station
	Start	Finish	(MHz)		(dBW)	called/worked

RSGB - Working for the Future of Amateur Radio

Station Log

Report		QSL		Remarks
sent	received	sent	rcvd	

Amateur Radio

Date	Time (UTC)		Frequency	Mode	Power	Station
	Start	Finish	(MHz)		(dBW)	called/worked

Station Log

Report		QSL		Remarks
sent	received	sent	rcvd	

Amateur Radio

Date	Time (UTC)		Frequency	Mode	Power	Station
	Start	Finish	(MHz)		(dBW)	called/worked

Station Log

| Report | | QSL | | Remarks |
sent	received	sent	rcvd	

Amateur Radio

Date	Time (UTC)		Frequency (MHz)	Mode	Power (dBW)	Station called/worked
	Start	Finish				

RSGB - Working for the Future of Amateur Radio

Station Log

Report		QSL		Remarks
sent	received	sent	rcvd	

Amateur Radio

Date	Time (UTC)		Frequency (MHz)	Mode	Power (dBW)	Station called/worked
	Start	Finish				

RSGB - Working for the Future of Amateur Radio

Station Log

Report		QSL		Remarks
sent	received	sent	rcvd	

Amateur Radio

Date	Time (UTC)		Frequency	Mode	Power	Station
	Start	Finish	(MHz)		(dBW)	called/worked

RSGB - Working for the Future of Amateur Radio

Station Log

Report		QSL		Remarks
sent	received	sent	rcvd	

WWW.RSGB.ORG   Tel: 01234 832700 Fax: 01234 831496

Amateur Radio

Date	Time (UTC)		Frequency	Mode	Power	Station
	Start	Finish	(MHz)		(dBW)	called/worked

RSGB - Working for the Future of Amateur Radio

Station Log

Report		QSL		Remarks
sent	received	sent	rcvd	

Amateur Radio

Date	Time (UTC)		Frequency (MHz)	Mode	Power (dBW)	Station called/worked
	Start	Finish				

RSGB - Working for the Future of Amateur Radio

Station Log

Report		QSL		Remarks
sent	received	sent	rcvd	

Amateur Radio

Date	Time (UTC)		Frequency (MHz)	Mode	Power (dBW)	Station called/worked
	Start	Finish				

RSGB - Working for the Future of Amateur Radio

Station Log

Report		QSL		Remarks
sent	received	sent	rcvd	

Abbreviations & Codes

Numerous abbreviations and codes are used by radio amateurs, especially when using Morse or datamodes. Many cross language barriers, enabling people without a common language to communicate, but some also find their way into spoken conversations when it would be just as simple to use plain language. Listed here are many of the common ones, but this list is by no means exhaustive and new abbreviations are being introduced all the time.

For a non-exhaustive list of acronyms, abbreviations and conventions as used in the Yearbook, RadCom and other RSGB publications please go to the RSGB website: *http://rsgb.org/main/publications-archives/radcom/supplementary-information/abbreviations-and-acronyms/*

Abbreviations

ABT	About
AGN	Again
AM	Amplitude Modulation
ANI	Any
ANT	Antenna
BCI	Broadcast Interference
BCNU	Be Seeing You
BK	Break
BTW	By The Way
BURO	(QSL) Bureau
B4	Before
CCT	Circuit
CFM	Confirm
CHK	Check
CLD	Called
CLG	Calling
CONDX	Conditions
CPI	Copy
CPY	Copy
CQ	General call ('Seek You')
CS	Call Sign
CU	See You
CU AGN	See You Again
CUD	Could
CUL	See You Later
CUZ	Because
CW	Continuous Wave (Morse)
DE	From
DN	Down
DR	Dear
DX	Long Distance
EL	Element
ENUF	Enough
ES	And
EU	Europe
EVE	Evening
FB	Fine Business (excellent)
FER	For
FM	Frequency Modulation
FM	From
FONE	Phone (telephony)
FREQ	Frequency
GA	Go Ahead
GA	Good Afternoon
GB	Good Bye
GD	Good
GD	Good Day
GE	Good Evening
GG	Going
GLD	Glad
GM	Good Morning
GN	Good Night
GND	Ground
GP	Ground Plane
GUD	Good

HI	Laughter
HI	High
HPE	Hope
HR	Here
HR	Hear
HRD	Heard
HV	Have
HVE	Have
HVG	Having
HVY	Heavy
HW	How
HW	How Copy?
IMI	Repeat (question mark)
INFO	Information
K	Invitation to Transmit
LID	A Bad Operator
LNG	Long
LP	Long Path
LSN	Listen
LTR	Later
LW	Long Wire
LW	Long Wave
MA	Millamperes
MGR	Manager
MI	My
MILS	Millamperes
MNI	Many
MOM	Moment
MSG	Message
MULT	Multiplier
N	No
NIL	Nothing
NR	Number
NR	Near
NW	Now
OB	Old Boy
OC	Old Chap
OK	Correct
OM	Old Man
OP	Operator
OT	Old Timer
OW	Old Woman
PLS	Please
PSE	Please
PWR	Power
R	Received
R	Are
RC	Ragchew
RCD	Received
RCVR	Receiver
RE	Regarding
REF	Referring to
RFI	Radio Frequency Interference
RIG	Station Equipment
RPRT	Report

RPT	Repeat
RPT	Report
RTTY	Radio Teletype
RST	Readability, Strength, Tone (report)
RX	Receive
RX	Receiver
SA	Say
SED	Said
SEZ	Says
SHUD	Should
SIG	Signal
SK	Silent Key (deceased)
SKED	Schedule
SN	Soon
SP	Short Path
SRI	Sorry
SSB	Single Side Band
STN	Station
SUM	Some
SWL	Short Wave Listener
TEMP	Temperature
TEST	Testing
TEST	Contest
THRU	Through
TKS	Thanks
TMW	Tomorrow
TNX	Thanks
TR	Transmit
T/R	Transmit/Receive
TRBL	Trouble
TRX	Transceiver
TT	That
TU	Thank You
TVI	Television Interference
TX	Transmitter; Transmit
U	You
UFB	Ultra Fine Business
UR	Your
UR	You're
URS	Yours
VERT	Vertical
VFB	Very Fine Business
VFO	Variable Frequency Oscillator
VY	Very
W	Watts
WID	With
WKD	Worked
WKG	Working
WL	Well
WL	Will
WPM	Words Per Minute
WRD	Word
WRK	Work
WUD	Would
WX	Weather

XCVR	Transceiver
XMAS	Christmas
XMTR	Transmitter
XTAL	Crystal
XYL	Wife
YF	Wife
YL	Young Lady (girlfriend)
YR	Year
Z	Zulu (Time)
55	Best Success
73	Best Regards
88	Love and Kisses

Five-unit Code

The so-called Murray Code is used by RTTY operators and telex machines. It consists of one 'start' bit, five 'data' bits, then 1.5 'stop' bits. As there are only 32 possible combinations of code, a limited character set is available (e.g. no lower case). Traditionally, amateur RTTY is sent at 45.45 Bauds, which results in an element length of 22ms. It is known officially as the International Telegraphic Alphabet No.2.

Binary	Dec	Hex	Octal	Letter	Figure
00000	0	00	00	[Blank]	
00001	1	01	01	T	5
00010	2	02	02	[Carr. Return]	
00011	3	03	03	O	9
00100	4	04	04	[Space]	
00101	5	05	05	H	# (note)
00110	6	06	06	N	,
00111	7	07	07	M	.
01000	8	08	10	[Line Feed]	
01001	9	09	11	L	)
01010	10	0A	12	R	4
01011	11	0B	13	G	& (note)
01100	12	0C	14	I	8
01101	13	0D	15	P	0
01110	14	0E	16	C	:
01111	15	0F	17	V	;
10000	16	10	20	E	3
10001	17	11	21	Z	"
10010	18	12	22	D	$
10011	19	13	23	B	?
10100	20	14	24	S	Bell
10101	21	15	25	Y	6
10110	22	16	26	F	! (note)
10111	23	17	27	X	/
11000	24	18	30	A	-
11001	25	19	31	W	2
11010	26	1A	32	J	'
11011	27	1B	33	[Figure Shift]	
11100	28	1C	34	U	7
11101	29	1D	35	Q	1
11110	30	1E	36	K	(
11111	31	1F	37	[Letter Shift]	

Note:
The letters F, G and H in the Figures mode are not allocated internationally. Each country is free to use them as they see fit. The American usage is shown in the table above.

Morse Code

The standard alphabet and numbers, as required for the Foundation Licence Morse Assessment.

A	·—		N	—·
B	—···		O	———
C	—·—·		P	·——·
D	—··		Q	——·—
E	·		R	·—·
F	··—·		S	···
G	——·		T	—
H	····		U	··—
I	··		V	···—
J	·———		W	·——
K	—·—		X	—··—
L	·—··		Y	—·——
M	——		Z	——··

1	·————		6	—····
2	··———		7	——···
3	···——		8	———··
4	····—		9	————·
5	·····		0	—————

Special Morse Characters

There are numerous procedural and punctuation characters. The following are those most commonly used by Morse operators.

A̅R̅ (+)	[di-dah-di-dah-dit]	End of message (used before the final calls and is written as 'AR' or '+')
C̅T̅	[dah-di-dah-di-dah]	Preliminary call
B̅T̅ (=)	[dah-di-di-di-dah]	Separation signal (used in text and is written as 'BT' or '=')
K̅N̅	[dah-di-dah-dah-dit]	Transmit only the station called (used after the final calls and is written as 'KN')
V̅A̅	[di-di-di-dah-di-dah]	Transmission ends (written as 'VA' or 'SK')
?	[di-di-dah-dah-di-dit]	Question mark (written as 'IMI' or '?')
/	[dah-di-di-dah-dit]	Oblique stroke (can be used as part of a callsign and is written as '/')
.	[di-dah-di-dah-di-dah]	Full stop
Error	[di-di-di-di-di-di-di-dit]	Erases the word in which a mistake has been made
@	[di-dah-dah-di-dah-dit]	As used in Email addresses

Phonetic Alphabet

The Phonetic Alphabet used by radio amateurs today was developed by NATO in the 1950s to be intelligible (and pronounceable) to all NATO allies. It replaced several other phonetic alphabets and is now widely used in business and telecommunications across Europe and North America.

A	Alpha	N	November
B	Bravo	O	Oscar
C	Charlie	P	Papa
D	Delta	Q	Quebec
E	Echo	R	Romeo
F	Foxtrot	S	Sierra
G	Golf	T	Tango
H	Hotel	U	Uniform
I	India	V	Victor
J	Juliet	W	Whiskey
K	Kilo	X	X-ray
L	Lima	Y	Yankee
M	Mike	Z	Zulu

Q Codes

There are a huge number of three-letter Q codes. Radio amateurs use only a small percentage of them, as many are of relevance only to shipping, aircraft, the police etc. They fall into the following pattern.

QAA-QNZ	Aeronautical
QOA-QQZ	Maritime
QRA-QUZ	All services
QZA-QZZ	Other

All Q codes follow the form of a question and answer. The list below gives details of those in common use by radio amateurs.

ASCII Code

The American Standard Code for Information Interchange is the code used in computing and for packet radio. Standard ASCII consists of seven data bits, which gives 128 possible combinations. The first 32 characters are used for control and the remaining 96 are representable characters.

Dec	Hex	Chr	Ctrl	Dec	Hex	Chr	Dec	Hex	Chr	Dec	Hex	Chr
0	0	NUL	^@	32	20	SP	64	40	@	96	60	`
1	1	SOH	^A	33	21	!	65	41	A	97	61	a
2	2	STX	^B	34	22	"	66	42	B	98	62	b
3	3	ETX	^C	35	23	#	67	43	C	99	63	c
4	4	EOT	^D	36	24	$	68	44	D	100	64	d
5	5	ENQ	^E	37	25	%	69	45	E	101	65	e
6	6	ACK	^F	38	26	&	70	46	F	102	66	f
7	7	BEL	^G	39	27	'	71	47	G	103	67	g
8	8	BS	^H	40	28	(	72	48	H	104	68	h
9	9	HT	^I	41	29	)	73	49	I	105	69	i
10	0A	LF	^J	42	2A	*	74	4A	J	106	6A	j
11	0B	VT	^K	43	2B	+	75	4B	K	107	6B	k
12	0C	FF	^L	44	2C	,	76	4C	L	108	6C	l
13	0D	CR	^M	45	2D	-	77	4D	M	109	6D	m
14	0E	SO	^N	46	2E	.	78	4E	N	100	6E	n
15	0F	SI	^O	47	2F	/	79	4F	O	111	6F	o
16	10	DLE	^P	48	30	0	80	50	P	112	70	p
17	11	DC1	^Q	49	31	1	81	51	Q	113	71	q
18	12	DC2	^R	50	32	2	82	52	R	114	72	r
19	13	DC3	^S	51	33	3	83	53	S	115	73	s
20	14	DC4	^T	52	34	4	84	54	T	116	74	t
21	15	NAK	^U	53	35	5	85	55	U	117	75	u
22	16	SYN	^V	54	36	6	86	56	V	118	76	v
23	17	ETB	^W	55	37	7	87	57	W	119	77	w
24	18	CAN	^X	56	38	8	88	58	X	120	78	x
25	19	EM	^Y	57	39	9	89	59	Y	121	79	y
26	1A	SUB	^Z	58	3A	:	90	5A	Z	122	7A	z
27	1B	ESC		59	3B	;	91	5B	[	123	7B	{
28	1C	FS		60	3C	<	92	5C	\	124	7C		
29	1D	GS		61	3D	=	93	5D	]	125	7D	}
30	1E	RS		62	3E	>	94	5E	^	126	7E	~
31	1F	US		63	3F	?	95	5F	_	127	7F	DEL

Q-code	Question	Answer	Colloquial use (if different)/explanation
QRB	How far are you from my station?	The distance between our stations is ...	
QRG	What is my exact frequency?	Your frequency is ...	Frequency of operation
QRH	Does my frequency vary?	Your frequency varies	
QRI	How is the tone of my transmission?	The tone of your transmission is ...	
QRL	Are you (or is the frequency) busy?	I am (or the frequency is) busy	
QRM	Are you suffering interference?	I am suffering interference	Man-made interference
QRN	Are you troubled by static?	I am troubled by static	Natural interference (atmospherics)
QRO	Shall I increase power?	Increase power	High power
QRP	Shall I decrease power?	Decrease power	Low power
QRQ	Shall I send faster?	Send faster	High speed
QRS	Shall I send more slowly?	Send slower	Low speed
QRT	Shall I stop transmitting?	Stop transmitting	To close down
QRU	Have you anything for me?	I have nothing for you	
QRV	Are you ready to transmit?	I am ready to transmit	
QRX	When will you call again?	I will call you again at ...	To stand-by
QRZ	Who is calling me?	You are being called by ...	
QSA	What is the strength of my signal?	The strength of your signal is ...	
QSB	Does the strength of my signals vary?	The strength of your signals varies	Fading
QSD	Is my keying defective?	Your keying is defective	
QSK	Can you hear me between your signals (and if so can I break-in)?	I can hear you between my signals (and it is OK to break-in on my transmission)	Break-in Morse operation
QSL	Can you acknowledge receipt?	I am acknowledging receipt	A card to confirm contact
QSO	Can you communicate with ... ?	I can communicate with ...	A contact
QSP	Will you relay to ...?	I will relay to ...	
QSX	Will you listen for ... (callsign) on ...?	I am listening for ... on ...	Split frequency operation
QSY	Shall I change frequency?	Change frequency to ...	
QTF	Will you give me the position of my station according to bearings taken?	The position of your station according to bearings taken is...	Beam heading (in degrees)
QTH	What is your position (location)?	My position (location) is ...	
QTR	What is the exact time?	The exact time is ...	